To my sons,
Birk and Bjorn

S.S.

Stone Ridge Press
2515 Garthus Road
Wrenshall, MN 55797
www.StoneRidgePress.com
thesparkygroup@gmail.com

BIRD NERD NATURAL HISTORY
WINTER FINCHES & FRIENDS OF NORTH AMERICA:
A NATURALIST'S HANDBOOK

Printed in South Korea by Doosan
10 9 8 7 6 5 4 3 2 1   First Edition

Graphic Designer: Mark Sparky Stensaas
Maps by Billy Anderson of Anderson's Art Farm (www.BillyAndersonsArt.com)
Base map by Matt Kania (www.MapHero.com)
Cover Photos by Sparky Stensaas (www.ThePhotoNaturalist.com)

ISBN-13: 978-0-9909158-1-2 softcover

Winter Finches

& Friends

of North America

a naturalist's handbook

by Sparky Stensaas

Stone Ridge
Press

Table of Contents

Map Key

Breeding

Winter

Year-round

Southern limit during
major irruptions

Sidebar Essays

Finch Fascination

Stepping out from the tiny cabin on that winter morning, I was greeted with an unreal cacophony of trills. It was December of 1987 and I was staying with a friend in the picturesque coastal town of Grand Marais, Minnesota on the rocky shore of Lake Superior just south of the Canadian border. The noise had penetrated the walls of the house but my mind couldn't imagine what was making the sound. The frigid air slapped me awake and I instantly became mesmerized by the sight and sound of several THOUSAND Bohemian Waxwings calling from the treetops. I did a slow turn and began estimating the numbers surrounding me in the hillside neighborhood. They weren't here yesterday, but today they were an army. Over the next week they efficiently stripped the entire town of its crop of crabapples and mountainash berries. And then they were gone, moving on to

White-winged Crossbill

find "greener pastures" which, to a Bohemian Waxwing means areas with bumper crops of sugary fruits.

The following March, at home in Duluth, another impressive congregation of birds stopped me in my tracks. This time it was a massive flock of Pine Siskins that had dropped in to feed on the seeds of the birches surrounding Rock Pond. Hundreds buzzed and trilled from the trees just above my head. I ran back inside and grabbed my parabolic reflector and a cassette tape recorder and tried to capture the sound experience. Impossible. Unforgettable.

Irruptive by Nature

Maybe it's the ephemeral nature of these winter visitors that intrigues me most (and many of us). That they can grace us with their presence one day, one week, one winter, then vanish the next. Unpredictable, because they are not on a normal migration, they are on a survival quest, searching out food sources that will get them through the winter.

Evening Grosbeak (L) and Red Crossbill (R).
[All from author's collection of Arm & Hammer Baking Soda Trading Cards]

Irruptive, is the name biologists give to the birds that "follow the food," and this is true for almost all the species in this book. Each winter, many of us can look forward to a new cast of avian characters.

Boreal Vagabonds

I am in love with all things boreal, I even love the word boreal. The North has always fascinated me and winter is arguably my favorite season. Growing up in Minnesota may have had something to do with it, as well as my Norwegian and Swedish ancestry.

L-R: Common Redpoll, Purple Finch, Pine Grosbeak.

Therefore, I am intrigued by birds that live in the vast and lonely boreal forest and Arctic tundra. Their trusting nature may be due to the species' infrequent contact with humanity. Several species are circumboreal, being found around the globe in northern forests (Pine Grosbeak, Common Redpoll, Hoary Redpoll, White-winged Crossbill, Red Crossbill and Bohemian Waxwing). I imagine it is the homogenity of the habitats, spruces, tamarack and tundra, that ring the globe from Saskatchewan to Scandinavia to Siberia.

Winter Finches AND Friends

Some of you may grimace at my eclectic choice of species to include in a book titled *Winter Finches*, but it is the winter wanderings that unites them in my mind. Snow Buntings, Bohemian Waxwings, Cedar Waxwings though not finches, all are songbirds that wander the winter landscapes of the northern U.S. and Canada. They and the Northern Cardinal round out the "& *Friends*" part of the book's title.

Thanks for joining me on this journey into the fascinating lives of our winter finches and friends.

—Sparky Stensaas

Nemadji Valley in northern Minnesota
March 2015

Snow Bunting

3

Bohemian Waxwing

Description Colorful and gregarious songbird that only graces the Lower 48 in winter, where it wanders in search of sugary fruits.

Length 8.25 inches (Starling-sized)

Other Names *Bombycilla garrulus* (Scientific), Waxwing (Europe), *jaseur boreal* (French Canadian), *Sidensvans* (Swedish), *Pestvogel* (Dutch), *Seidenschwanz* ("Silky Tail" in German), *Silkitoppa* ("Silky Crest" in Icelandic), *Tilhi* (Finnish)

Hot Spots Winter—Residential Sault Ste Marie, Michigan; St. John's, Newfoundland; Minnesota's North Shore of Lake Superior from Duluth to Grand Marais.
Early Spring—Golden Gate Canyon State Park in Colorado.
Summer—Alaska's Denali National Park, Chugach National Forest near Anchorage; Churchill, Manitoba.

Finding a flock of Bohemians is always a happy event. Even if I'm on my way to a board meeting or my kid's ski practice, I'll stop and gawk at a flock as they work over the fruits on a crabapple or mountain-ash. This species' erratic winter wanderings are often given as the reason for them being tagged with the moniker, "Bohemian." But wouldn't "Gypsy Waxwing" be a more fitting name?

Description & ID Tips

Similar to the more familiar Cedar Waxwing but larger, grayer and more colorful. The previous sentence sounds a bit contradictory, but their body is more gray (compared to the Cedar's warm brown and yellow coloration) and they show additional color on the wings (yellow and white edging in addition to the red "wax" tips). Face, forehead and under tail coverts are a rusty red-orange (Cedars have white feathers under their tail). And under the tail is the first place I look when trying to identify a Bohemian. Females are very similar to males but tend to have a smaller black chin patch, more diffuse border to the gray breast and fewer and shorter red wax tips to secondaries.

Resemble starlings in flight silhouette. Fly in tight formation.

Songs and Calls

Their husky trill (deeper and louder than Cedar Waxwing's call) has several subtle variations likely used in communicating to each other. It is not a true song. Since their primary foods are abundant and ephemeral fruits, which they could not defend even if they wanted to, they are monogamous and have little use for a true song.

Please Pass the Crabapple

Courtship in the waxwings is a charming affair. Male hops sideways along a branch to sidle up to a potential mate. He tries to touch her bill with his. If she responds with her bill, the connection is made. This behavior may also include "gift passing" whereby the male passes a fruit, insect, flower petal or even an inedible object to the female. She may respond by taking the gift, hopping sideways away from the male then returning to him and passing it back. The whole performance may be repeated a dozen times before she finally eats the object and seals the deal.

Courtship in waxwings includes the male sidling up to the female, and sometimes offering her fruit.

2

Habitat

Breeding—Open boreal spruce forest and mixed conifer-birch woods. Proximity to water and open habitat with flycatching perches is important.

Winter—Anywhere in the northern tier of states and Interior West where crabapples, buckthorn, mountain-ash and other sugary fruits are abundant.

Range

Breeding—Taiga of central Alaska across Canada to south shore of Hudson Bay in northwest Ontario, including northern Manitoba, northern Saskatchewan, northern Alberta, interior British Columbia, Yukon and Northwest Territories. Very rare breeder to Montana, Washington, Idaho and Oregon. Circumboreal; across Scandinavia and Siberia.

Winter—Highly irruptive species. Some winters they are non-existant in northern Minnesota, and other years there are many thousands. I vividly remember stepping out of a friend's house in Grand Marais, Minnesota one December day and being knocked back on my heels by the cacophony of waxwing calls. I estimated 2,000 were visible from one spot!

Nest

Tree-nester that prefers conifers on the edge of an opening or near a lake, pond or stream. Both the male and female quickly (3 to 5 days) build the nest of pine and tamarack twigs and grass. They camouflage the outside with Reindeer Lichen (*Cladonia* spp.) and mosses, and line it with down, cocoons and

3

2. Bohemians make forays to the ground to eat fallen fruit and "drink" snow.

3. Ornamental crabapple trees have been a boon to waxwings and many other fruit-eating birds.

hair. Nest appears disheveled with plants hanging from it. In areas with abundant fruiting trees, many pairs may nest in close proximity, taking advantage of the nearby food source. Normally placed five to 20 feet up in the tree. Two nests found near Churchill, Manitoba were about ten feet up in a Black Spruce but made with tamarack branchlets. Five eggs is the norm, four to six sometimes.

Food

Sugary fruits are in high demand in winter, especially crabapples, and fruits of mountain-ash, at least in the North Woods. Highbush cranberry, buckthorn, Russian olive, rose hips, cranberries, grapes, hawthorn fruits are also eaten. Dried winter fruits must make them thirsty, as you often see them dropping down to the ground to take in snow or water. Seeds of

4. Flocks often stage in the high tops of trees near a fruit food source.

Bohemian or Cedar?

1. UNDERTAIL COVERTS
Bohemian—Rusty or rufous
Cedar—White

2. BODY COLOR
Bohemian—Gray
Cedar—Warm brown

3. WING MARKINGS
Bohemian—Yellow, white, red
Cedar—Only red "wax" tips

4. HEAD & FACE COLORS
Bohemian—Rufous forehead and cheeks
Cedar—White-edged mask

5. CALL
Bohemian—Husky trill, deeper than Cedar's
Cedar—Thinner, softer trill

Cedar

Bohemian

Drunk & Disorderly

Bohemian and Cedar Waxwings can get quite drunk on fermented fruit. In a case of avian forensics by Stephen and Walley (2000), this was clearly proven by chemical analysis. The victims (16 Bohemian Waxwings and a Pine Grosbeak) were killed when they crashed into a building near a crabapple tree in Dauphin, Manitoba. The researchers analyzed the partly digested apple pulp in the birds' gizzards and discovered an alcohol content near 3 percent and a blood alcohol level of 73 mg per 100 ml—for humans, this is near or above the legal limit for driving, let alone flying!

Though Bohemians can metabolize alcohol faster than Cedar Waxwings, Cedars are also vulnerable to intoxication. In one California case the mass mortality of 42 Cedars was attributed to eating fermented palm fruits.

Bohemians often have to perform acrobatic maneuvers to get at mountain-ash berries.

the host fruit trees are usually defecated out unharmed, and ready to germinate.

In early spring, Bohemians seek a sweet snack by sipping maple sap that oozes from cracks, wounds and broken maple twigs. Hovering below "sapsicles," waxwings catch drips of melting sap. Bohemians, like their Cedar cousins, hawk and sally for flying insects during the warm months.

Nature Notes

Large livers allow Bohemians to process ethanol from fermented winter fruits better than most birds.

Breed late so tree fruits are ripe when nestlings need to be fed.

Salt poisoning may be an understudied cause of mortality in frugivores like Bohemian Waxwings. When they drink from puddles that have a high concentration of dissolved road salt, their coordination may be impaired and suffer mortality from cars and hitting windows.

There are three waxwings in world. The only one not found in North America is the Japanese Waxwing (B. japonica), which has a red-tipped tail and red wing stripe with black and white barring on wings. Breeds in the coniferous forests of Far East Russia and China but winters in Japan and Korea.

Badge of Status

Scan to see video of Bohemians feeding on fruit in the wild.

The red waxlike tips on the adult's secondary wing feathers, aren't just static plumage, but rather increase in number and size as the bird ages, and play a role in social stature. The more mature the waxwing, the more young it is able to fledge on average. The white wing crescent and yellow tail band also increase in size and brightness in older individuals.

5. By late winter smaller groups merge into "mega flocks," sometimes numbering in the hundreds or even the thousands.

Cedar Waxwing

Description Crested and masked songbird that feasts on sugary fruits in summer and winter. Sometimes flycatches for aerial insects.

Length 7.25 inches

Other Names *Bombycilla cedrorum* (Scientific), *jaseur d'Amerique* (French Canadian), cankerbird (fondness for cankerworms)

Hot Spots Widespread.

Popular and widespread songbird across North America. Intimately tied to crops of sugary fruits including mountain-ash. Increasing in population and range. In the North Woods it often moves south in fall to be replaced by irruptive winter Bohemian Waxwings arriving from the north.

Description & ID Tips

Crested warm-brown bird with yellowish belly and black facemask. Red "wax" tips on secondary flight feathers of the wings. Noticeably smaller than its Bohemian cousin and also lacks rusty red undertail coverts (whitish in Cedar), gray belly (yellowish in Cedar) and white and yellow wing stripes (none on Cedar).

Song and Calls

Waxwings do not sing and the high thin trills and whistles we associate with them are actually calls. But these calls are subtly varied and can communicate many things to their companions and flock mates.

Habitat

Breeding—Wide variety of habitats from woods to orchards, but always with edge habitat and often near water. Avoids the deep dark interior of woods. Less site tenacity to former nesting areas due to need for good fruit crop to feed young.

Winter—Erratic range due to variable availability of fruit crops. May be common in an area one winter, but absent the next.

Range

Breeding—Central Canada south to Georgia, Arkansas, Colorado, Utah, Nevada and California. In Alaska only breeds in the far southeast panhandle.

---

1. Mountain-ash (*Sorbus* sp.) fruits are highly sought in late summer through early spring.

Fatty Fruit?

Both Cedars and Bohemians can survive and thrive on sugary fruits for long periods of time. Mountain-ash fruits are a huge source of winter energy for Bohemians. Studies by Pulliainen (1978) show that they are able to store excess assimilated sugars as fat, a good thing on a minus 30 F night! Both species also feed on sucrose-rich foods including sap, buds, flowers and scale insects. For many birds, sucrose is difficult to digest, but Cedars have a trick "up their intestines"—sucrose. This enzyme allows them to break sucrase into the more easily digested glucose and fructose.

Juvenile Cedar Waxwings feasting on mountain-ash fruits in August.

Winter—Winters from southern Canada to Panama but most common in southern 2/3 of the U.S. becoming much less common through Mexico and rare in Costa Rica and Panama. Some winter in the Caribbean as well.

Nest

Tree-nester that builds a bulky cup nest of twigs, dry grasses and sometimes *Usnea* lichens and lined with pine needles, plant down, wool, rootlets and fine grasses. Nest is frequently placed well out on a branch and from five feet up to much higher, but usually within 12 feet of the ground. Clutch ranges from three to six eggs with four to five being most common.

Food

Heavily reliant on sugary fruits in winter, especially those high in easily digestible sugars such as glucose and fructose. Cedar berries have always been an important fruit, even giving this bird its common name. In northern parts of its range mountain-ash is preferred. Crabapple, juneberry, chokecherry, pyra-

cantha, privet, mulberry, blueberries, chokecherries, saskatoons, and juniper berries are all choice edibles. In southwest U.S. eats toyon, mistletoe, madrone, juniper and peppertree fruits.

Some things never change; John James Audubon commented in 1842 that "the appetite of the cedar-bird is of so extraordinary a nature as to prompt it to devour

3. Cedar Waxwing pair.

4. Accessing hanging fruits sometimes requires acrobatics.

every fruit or berry that comes its way." They will occasionally gorge themselves until they cannot get airborne.

Ornamental plantings in yards and parks have proved a boon to waxwings who freely help themselves to crabapples, hawthorns, Russian-olive, firethorn and cultivated forms of mountain-ash. Sadly they also feed on alien buckthorn berries, spreading the aggressive invasive wherever they poop.

Spring diet may include petals of apple blossoms. Waxwings sally from exposed perches, flycatching aerial bugs, and these flying insects in summer are a very important protein source. Red Osier Dogwood berries are often fed to nestlings as a mash.

Nature Notes

Their curious common name "waxwing" comes from the red tips on their wing feathers that to some early naturalist resembled dabs of red wax used to seal envelopes in "days of yore."

Adults and young "freeze" in position when alarmed.

Adults feed young a sticky white mash called "pap" which they regurgitate from their gullet. This is likely the pulp of tree fruits.

Late Nester

Cedar Waxwings are one of the latest nesters in North America with egg-laying initiated from July to early August. And for good reason, as their important fruit food source is peaking later in the summer. Fledglings seen as late as August 27th in Manitoba. Loosely colonial breeders (8 nests in North Carolina clustered in a very small area) with a high tolerance for close neighbors; a Cedar Waxwing nested within two feet of a pair of Chipping Sparrows in Pinawa, Manitoba. But a cowbird egg in their nest is another matter, and once discovered, it will be ejected quickly.

Juvenile Cedar Waxwing.

Northern Cardinal

Description Common crested red bird of yards, parks, gardens and many other habitats with dense cover. Iconic and much-loved, but often shy and retiring.

Length 8.75 inches (Cardinal-sized!)

Other Names *Cardinalis cardinalis* (Scientific), *cardinal rouge* (French Canadian)

Hot Spots Widespread.

W ho doesn't love the cardinal? (Except those trying to sleep in on a Sunday morning?) Cardinals are the red and crested ambassador for all birds. They are the poster child for Christmas, emblazoned on many a Holiday card. Though not a finch, I included them because they enliven many of our winter feeding stations.

Description & ID Tips
Unmistakable red bird with black face and chin, prominent crest and heavy red bill (juveniles have black bill). Females are dull red and warm brown. Females may be confused with Pyrrhuloxias but note that the Cardinal always has a red bill and the Pyrrhuloxia always has a yellow bill.

Song and Calls
Loud series of clear whistled *whitchew whitchew ... whit whit whit ... tchew tchew tchew* notes given in different combinations. Often sings predawn and dusk, but can continue through the day.

Habitat
Year-round—Habitat generalist: parks, suburbs, towns, forest edges, successional fields, margins of ponds and streams. Anywhere small trees and dense cover present.

Range
Year-round—Slowly moving north and west, likely due to increased winter bird feeding and more edge habitat. Northern Minnesota and Maine south to Florida, Texas and Arizona. South of the border through Mexico to Guatemala.

Nest
Nest tucked deep into thick vegetation including vines,

Headin' North

The Cardinal's range expansion north is not a recent phenomenon. Their steady march has been going on at least since the mid 1800s! Three factors may be responsible for the movement. 1) Clearing of the land and the resulting second growth and increased edge habitat. 2) Growth in popularity of winter bird feeders. 3) Possibly climate change, though this is speculation.

First recorded in Minnesota in 1890s, they moved north to Minneapolis by 1930 and are just now becoming established in Duluth. It has taken the Cardinal over 100 years to move 250 miles north.

The Cardinal is slowly moving north and west.

shrubs, and small coniferous trees, usually below ten feet. Loosely woven cup of grasses, twigs, vines, bark strips that is lined with finer grasses and hair. Three to four eggs normally, occasionally two or five.

Food

Adults eat mostly fruits (mulberry, sumac, dogwood, etc.) and seeds (sedges, smartweeds, etc.). Heavy bill is adept at crushing and extracting seeds. Nestlings are fed exclusively animal matter—insects, spiders, centipedes.

Nature Notes

Able to deftly peel the skin off wild grapes to get at the pulp and seeds.

One of the first birds to the feeders in the morning (often pre-sunrise) and one of the last to depart.

Oldest Cardinal was at least 15 years, 9 months old.

State Bird of six states: IL, IN, KY, NC, OH, VA, WV.

Southern Cousin

Pyrrhuloxias could easily be confused with female cardinals...If you happened to be in Texas, New Mexico or Arizona, the only places where the two species overlap. They are sympatric in the southwest U.S. with identical breeding cycles, overlapping territories and even similar songs, but no aggression between the species has been noted. Compare Cardinal's red pointed bill to Pyrrhuloxia's curved yellow bill.

Cardinal female

Pyrrhuloxia

Thrushes in Winter

A variety of thrushes can enliven our backyard fruit trees in winter. American Robins may linger in the north if there are enough fruit-laden mountain-ash or crabapple trees in the neighborhood. The males return early in spring to set up territories, and since the ground is often still frozen, worms and other invertebrates may not be available, so they instead feast on the frozen fruits.

If you live in the eastern U.S., a sighting of one of the western thrushes will surely make your day, and probably make it onto the local rare bird alert. Townsend's Solitaires and Varied Thrushes are natives to the Pacific Northwest, British Columbia, the Yukon, Alaska and the Interior West (Townsend's only). They routinely wander east to the Midwest, Great Lakes and Northeast U.S.

I had to include the Fieldfare as well. A European thrush that has only been recorded a handful of times in North America. I photographed this one in November 1991 near Grand Marais, Minnesota, just south of the Canadian border. It was the first state record for Minnesota and the furthest west ever recorded in the Lower 48.

Clockwise from top left: American Robin, male Varied Thrush, Townsend's Solitaire, Fieldfare.

Snow Bunting

Description Black-and-white winter visitor that forages in large flocks in open areas. Breeds in the High Arctic.

Length 6.75 inches

Other Names *Plectrophenax nivalis* (Scientific), *bruant des neiges* (French Canadian), *Snösparv* (Swedish), *Sneeuwgors* (Dutch), *Schneeammer* (German), *Snøspurv* (Norwegian), *Snjótittlingur* (Icelandic), *Pulmunen* (Finnish), *Qaulluqtaaq* (Inupiaq Eskimo), *Kó-ka-noch* (Yupik Eskimo)

Hot Spots Winter—Sax-Zim Bog in northern Minnesota, Aitkin County, Minnesota. Summer—Churchill, Manitoba (until early June); Gambell, Barrow, Nome and Homer Spit (all in Alaska).

Y ou know you better finish up all your outdoor chores around the house when you see the first flock of Snow Buntings in fall because winter is not far behind.

Rising from winter roadsides in a swirl of white and black, Snow Bunting flocks undulate over the barren landscape, flashing black and white, sometimes appearing to disappear. They descend on snowy fields where they scurry about feeding on weed seeds atop the snow, occasionally reaching up or hopping to snatch a seed. In big foraging flocks, sometimes the rear guard flies up and lands just ahead of the rest. When snows are deep, they often feed along snow-free road shoulders and railroad tracks where foraging is easy and spilled grain from trains and trucks, and windblown seeds collect.

Description & ID Tips
The white body and black-marked wings is what first catches our eye of this winter visitor. Most often in flocks along roadways which scatter at our oncoming car revealing bold black and white markings. Yellow bill in winter, black on breeding grounds.

Song and Calls
We usually only hear their call *tiriririt* followed by clear *pyu*. In its Arctic breeding grounds the males often sing from the highest thing around, which is usually a boulder. Like other species in open country with few perches, they also have an aerial display and song. They rise to about 30 feet then start singing as they return to terra firma on stiff fluttering wings.

Habitat
Breeding—Bare rocky areas: scree, sea cliffs, tundra.

1. Beginning to show its breeding plumage.
2. Female in fall/winter plumage. [Inset: Feeding on weed seeds]

Snow Roosting

Snow roosting is a well documented survival strategy for Ruffed Grouse and Sharp-tailed Grouse. At dusk the grouse burrow into deep snow and sleep about a foot below the surface where it may be up to 40 degrees warmer than the air temp.

I witnessed this flock of Snow Buntings hunker down in an open field in northern Minnesota's Sax-Zim Bog one winter day. Upon closer inspection, several had nestled deep in the snow so only half their body was exposed. They were not foraging and remained in this spot for awhile. This behavior is likely a method for retaining body heat and reducing exposure to cold winds. Though normally fine in the open, they are known to hide behind drifts and burrow into snow during extreme cold.

Flock of Snow Buntings "snow roosting" in an open field.

Winter—Anywhere snow cover is minimal and seeds can be found: roadsides, shoulders, railroad tracks and right-of-ways, farm fields, weedy meadows.

Range

Breeding—Nests farther north than any other passerine, mostly north of 68 degrees latitude. Circumpolar, ranging from Alaska east through the Canadian Arctic, Aleutians, Greenland, Iceland, Svalbard, Scotland, northern Scandinavia, Siberia and Kamchatka.

Migration—Some populations migrate southeast while others move southwest, and the two groups

3. Juvenile Snow Bunting (Iceland).   4. Drifts of Snow Buntings often forage in northern farm fields.

may even cross paths. Birds from western Greenland winter in North America while those from the other side of the island head southeast to the Russian steppe or the British Isles.

Winter—Southern Canada to northern tier of states. Rarely seen south to Texas, Arkansas, South Carolina. Males winter farther north than females.

Nest

Ground-nester that makes a moss, lichen and dry grass cup lined with fine grasses, hair, fur and feathers. Placement is usually in rock crevices (including holes in a wall or foundation, nook under a boulder or even the eave of a building). They occasionally re-use old nests. Sometimes communal with many pairs nesting in the same area. Four to six eggs.

5. Note the black bill of this summer-plumaged Snow Bunting.

6. Female Snow Bunting feeds drab-colored juvenile (Iceland).

White or Black?

Most birds acquire breeding plumage through a spring molt, but not the Snow Bunting. Females and winter males show a fair amount of buff and brown plumage, but the male's bold black and white plumage is only hidden. As you can see in the photo of the wind-blown bird below, the Snow Bunting's "white" feathers are mostly black (50-70 percent). The black on it's back only reveals itself as his feather tips wear off in late winter and spring. Adults actively attempt to abrade the buff and tan tips off their feathers by rubbing their head, breast, flanks, back and scapulars on the snow, and this is how they attain their beautiful snow-white-and-jet-black summer plumage.

Snow Bunting feathers are mostly black as you can see on the breast of this wind-blown bird.

A "Drift of Buntings"

Winter flocks in the far northern U.S. are usually "pure bunting," but a bit farther south they often host several guest Horned Larks or Lapland Longspurs. In the north-central U.S. this ratio may be reversed with Snow Buntings being in the minority.

Winged Blizzards

T. S. Roberts, the pioneering Minnesota bird expert and author of *The Birds of Minnesota*, recorded some massive flocks of Snow Buntings. Squaw Lake in Itasca County hosted a group of 2,500 to 3,000 in late October 1925, which he labeled "a small snowstorm." Some amazing spring migration counts north of the border dwarf these flocks. Over 18,000 were seen between Elm Creek and Oakville, Manitoba on April 30, 1996, and what can only be described as a winged blizzard occurred on April 19, 1993 when a staggering 50,000 were seen between Brunkild and Lowe Farm, Manitoba.

Mob Mentality

Falcons seem to create a mob mentality in these seemingly docile "snowflakes." And I guess it makes sense, as Merlins, Prairie Falcons and even Gyrfalcons will gladly pluck a Snow Bunting snack from a winter flock. Five hundred Snow Buntings mobbed a Merlin near Seven Sisters Falls, Manitoba and another observer witnessed an estimated ONE THOUSAND pursue two Prairie Falcons near Stead, Manitoba on November 6, 1988. That would be quite a sight!

Food

Seeds of grasses and weeds. Does not eat sunflower seeds at feeders.

Nature Notes

Vagrants have wandered as far afield as the Canary Islands, Morocco, Bulgaria, Turkey, Bermuda, Florida, Texas and Hawaii. Buntings banded in Michigan have showed up in Yakutat, Alaska.

Most amazing is the eyewitness account of renowned Norwegian Arctic explorer Fridjof Nansen who reported Snow Buntings four times between 84 and 85 degrees North latitude in May and June of 1895. This is 200-plus miles north of the nearest known *terra firma* (Franz Josef Land) and the farthest north any songbird has ever been seen.

7. Snow Buntings are rarely seen in trees.

8. You are most likely to see them foraging along winter roads.

Arctic Cousin
McKay's Bunting

Amazingly, this is Alaska's only endemic bird species! Its entire population of 3,000-6,000 birds live on Hall and St. Matthew's Island in the Bering Strait. Sometimes breeds on St. Lawrence Island and in the Pribilofs on St. Paul and St. George. Strays have appeared in British Columbia, Oregon and Washington state.

Migrates to sea coasts of Alaska and is rare to uncommon at feeders in western Alaska. Also possibly at Nome in March.

Named after Charles McKay of the U.S. Army who was a pioneer collector of Alaskan birds for the U.S. National Museum.

This wayward McKay's Bunting showed up in coastal Washington state in 2012.

Evening Grosbeak

Description Stunning black and yellow finch of the boreal forest and mountain conifers. Loud and gregarious. Big-headed and big billed.

Length 8 inches (Cardinal-sized)

Other Names *Coccothraustes vespertinus* (Scientific), *gros-bec errant* (French Canadian)

Hot Spots Winter—Feeders in Sax-Zim Bog and around Ely, Minnesota.
Early Spring—Golden Gate Canyon State Park in Colorado.
Summer—Moraine Park Campground in Rocky Mountain National Park; Cascades near Leavenworth, Washington.

1

A much-loved and boldly-marked finch of the North Woods and mountain forests. Its dramatic dips and booms in population are an endless source of speculation. They were abundant at backyard bird feeders across the northern U.S. in the 70s and 80s but have declined dramatically.

Description & ID Tips

Gros in French means "large," so grosbeak is an appropriate name for this heavy-billed finch. Yellow, black and white male is unmistakable, and female shows the same color scheme but in paler shades. I especially like how the black on the head transitions into yellow on the belly. Yellow "eyebrows" on the male look more like "Viking horns" to me. Plumage does not change seasonally. In flight their large white wing patches flash as they zip overhead. Large-headed and short-tailed.

Song and Calls

A songbird that doesn't regularly sing. Loud ringing and trilly *deer* and a sharp *tew* are call notes, and seem to take care of much of their communication needs.

Habitat

Boreal and other coniferous forests across North America.

Range

Breeding—Southeast corner of the Yukon across the boreal forests of Canada to northern New England, south to northern Michigan, Wisconsin, Minnesota, Rocky Mountains, Cascades, Sierra Nevadas and even to mountains of Mexico.

Winter—Same as breeding range except in years of major irruption when they may move south into the central and southern U.S.

1. The male Evening Grosbeak is unmistakable.

Budworm Boon

Prior to the 1850s, Evening Grosbeaks were unknown east of the western Great Lakes. One theory on how they exploded to the Northeast U.S. and Canada during the 1900s is tied to massive outbreaks of Eastern Spruce Budworm (*Choristoneura fumiferana*). While the insect is the bane of Balsam Fir and spruce forests, it is a boon to these black-and-yellow finches who devour them by the millions. Abundant food allowed them to lay more eggs, raise more young. Others credit plantings of Box Elder across the East; Its seeds are a grosbeak favorite.

Outbreaks of Spruce Budworm, a native pest, are a major food source for Evening Grosbeaks.

Paushkundamo

Evening Grosbeaks were first described in 1825 by the great naturalist William Cooper (think "Cooper's Hawk") from a specimen sent to him by Henry Rowe Schoolcraft (The first white person to see the source of the Mississippi River). It was shot by a native boy on the evening of April 7, 1823 near Sault Ste. Marie, Michigan Territory and labeled *Paushkundamo*, an Ojibwa word meaning "berry-breaker." He also noted that the species is "said to be common about the Head of Lake Superior at Fond du Lac" (present day Duluth, Minnesota and only a few miles from my home!)

Henry Rowe Schoolcraft

A few months later, along Minnesota's Savanna River, Major Delafield also encountered this new bird. His twilight observation probably helped give this beautiful bird its misbegotten common name (Evening) and Latin name (*vespertinus*). "*It's mournful cry about the hour of my encamping (which was at sunset) had before attracted my attention…[and] my inference was…that this bird dwells in such dark retreats and leaves them at the approach of night.*"

Nest

Tree-nester that places its flimsy nest rather high up, from 20 to 60 feet in a coniferous tree, often at a level between 60 and 80 percent of the tree's height. Loosely built of sticks and woven with moss and lichens into an oval flattened nest that is lined with rootlets, hair and fibers. So spare that sometimes the eggs can be seen through the bottom of the nest! Inconspicuous during the breeding season so very few nests are actually located. Three to four eggs, sometimes two or five.

Food

Box Elder seeds seem to be a favorite wild food, as do the keys of other maples species. Also seeks out seeds of sumac, ash, elm and pines, flowers of birch and maple, berries of crabapple, Russian olive, hawthorn and juniper. Of course, we usually see winter birds gorging on black oil sunflower seeds at bird feeders. Buds of maple, elm, oak, aspen, willow and cherry. In summer, insects are relished, especially the easily captured Spruce Budworms (see sidebar on page 25).

2. It is a real treat to host a flock of Evening Grosbeaks in winter.

3. Chester A. Reed illustration from a 1903 field guide.

Nature Notes

Oldest recorded free-flying wild Evening Grosbeak was an amazing 15 years and 3 months!

In spring they may snap off a maple twig knowing that doing so will create a sweet "spigot" of dripping maple sap allowing them a refreshing drink.

Like the crossbills, Evening Grosbeaks have a "salt tooth," craving the mineral to the point of foraging near and under cars dripping with road-salt-soaked snow.

Of 17,000 Evening Grosbeaks banded in Pennsylvania over 14 years, 499 were recovered. Only 10 percent were returnees to Pennsylvania, while the other 90 percent had wandered across 17 states and four provinces!

Why such a Heavy Beak?

Obviously Evening Grosbeaks don't need their massive bill for feeding on the soft and gooey spruce budworm caterpillars that they love so much. So why the big bill? The answer is for seed crushing. Unlike waxwings, who love the fleshy pulp of berries, these grosbeaks actually discard the pulp and skin and keep the seed. It is then positioned just right in the beak and upwards of 60-125 lbs. of force is brought down upon the seed to crack it. One observer watching an Evening Grosbeak eating wild cherries could hear the "pop" of a breaking pit 100 feet away! This skill comes in handy when feeding on the extremely hard cones of Bald Cypress in South Carolina and Florida. The now extinct Carolina Parakeet was a master at getting at the Cypress seeds, but very few other animals can.

Big beaks are effective at cracking hard seeds and fruit stones.

4. Females lack the bold colors of the males but they are still very attractive birds.

Pine Grosbeak

Description Elusive robin-sized bird of the far northern boreal forests and mountains. Female is olive-yellow but male is striking rosy-red. Winter visitor to sunflower seed feeders in northern tier of states.

Length 9 inches (Robin-sized)

Other Names *Pinicolor enucleator* (Scientific), *durbec des sapins* (French Canadian), *Tallbit* (Swedish), *Haakbek* (Dutch), *Hakengimpel* (German), *Konglebit* (Norwegian), *Krókfinka* ("Hook Finch" in Icelandic)

Hot Spots Winter—Northern Minnesota's Sax-Zim Bog; St. John's, Newfoundland.
Early Spring—Golden Gate Canyon State Park in Colorado.
Summer—Utah's Wasatch-Cache National Forest.

I really love the Pine's "tweedling" song. To me it says winter. In my deer hunting days, I'd often hear my first one of the fall while up in the stand. In northern Minnesota deer season is in early November and this is about when they'd make their first appearance.

Description & ID Tips

What color is the male Pine Grosbeak? Sibley says pinkish-red. Peterson said rose-red. Wine? Burgundy? Carmine? However you describe the color, it is an impressive bird—plump, stocky and long-tailed. Also note its prominent white wing bars and beautiful scalloped feathers on upper back (see main photo on facing page). Females are gray with greenish yellow head and rump.

Song and Calls

Musical and high-pitched tweedling warble. Relatively loud and sweet. They also have a winter "whisper song" that is long and quiet (can be heard to 100 feet or so), and, according to some "earwitnesses," sometimes interspersed with imitations of the calls of other boreal-nesting birds.

Habitat

Breeding—Subarctic and boreal forests that are rather open. Avoids dense coniferous forests and aspen stands. In Alaska

1. Adult male Pine Grosbeak feasts on crabapples.

Winter Finch Forecast

Each fall hardcore birders and bird guides look forward to a big event. Thanksgiving? No. Ron Pittaway's Winter Finch Forecast. Ron is a founding member of the Ontario Field Ornithologists and a former naturalist at Algonquin Park. He garners tree seed crop intelligence from across Canada and Alaska including the Ontario Ministry of Natural Resources, eBird Canada and many dedicated birder-naturalists. Why are birder's gawking at tree tops? Because for many of our boreal finches, movements and wanderings are all about food, and the key trees are spruces, birches and mountain-ash. Once the data is compiled, Ron works his prognosticating magic. While geared specifically to eastern Ontario, including Algonquin Park, the forecasts are often very useful for the northern tier of the U.S. as well. For example, the most recent winter he predicted a "bumper crop" of Purple Finches (ding, ding…correct!), a moderate to good flight of Common Redpolls (I would say in northern Minnesota we had a major irruption), small flight of Pine Grosbeaks (right again, Ron!), White-winged Crossbills would be mostly absent (absolutely accurate! I saw two all winter), He also makes predictions for Red Crossbills, Hoary Redpolls, Evening Grosbeaks and three irruptive songbirds—Blue Jays, Bohemian Waxwings, and Red-breasted Nuthatches.

Female or Russet Male?

What about those gray birds with the russet cap and rump? These are likely young males, but some could also be adult "russet females."

Juvenile males and females are mainly gray with golden-olive or orangey-russet crown and rump. Young males incrementally turn rosy red and in late winter we can see some interesting and beautiful color patterns on these birds, as in photo bottom left. This gorgeous bird shows a mosaic of pink, gold, orange, red and russet. It was photographed in late February on the North Shore of Lake Superior.

Female in crabapple (top right). Basic I plumage of male shows russet head and rump (bottom right), but some adult females also show this color form. Male transitioning to definitive basic plumage (bottom left).

2

may nest far from trees in areas with alder thickets. High, wet valleys near the treeline in the Rocky Mountains. Rare outside the boreal forest.

Winter—Anywhere there is abundant ash and maple seeds, mountain-ash berries and crabapple fruits (and sunflower seeds at feeders).

Range

Breeding—Circumboreal. Alaska across Canada to Scandinavia, Siberia, Kamchatka and Hokkaido in Japan. Dips down into the Lower 48 in the Rocky Mountains and the Cascades. Isolated populations in the White Mountains of east central Arizona and California's Sierra Nevadas.

Winter—Often remains in breeding areas but usually appears in modest numbers in southern Canada, northern New England and northern Minnesota, Wisconsin and Michigan. Can irrupt in larger numbers during some winters.

Nest

Loosely constructed twig nest placed low (two to ten feet up) in a spruce, fir, birch or juniper. Usually tucks it next to the trunk on the south side of tree. Inner nest is lined with moss, rootlets and grass. Four eggs is the norm, but two to five recorded.

Food

Favorites include maple and ash keys, birch seeds, mountain-ash fruit, crabapples, juniper berries, Highbush Cranberries, spruce cones and lilac seeds. Commonly eats sunflower seeds at feeders.

Nature Notes

Seems to be faithful to both wintering sites and breeding locations. Banded birds have been recaptured up to five years later at the same winter site.

Recordings of flight calls of wintering Minnesota grosbeaks indicate that the birds originated in the central taiga west of Hudson Bay.

3

2 & 3. Pine Grosbeaks are often found utilizing ornamental crabapples and bird feeders in their winter range.

Gray-crowned Rosy-Finch

Description Chunky songbird of the high mountains and tundra. Comes to lower elevations in mid-winter.

Length 6.25 inches (House-Sparrow-sized)

Other Names *Leucosticte tephrocotis* (Scientific), Gray-crowned Leucosticte (old name), Gray-cheeked Rosy-Finch, Hepburn's Rosy-Finch, *roselin a tete grise* (French Canadian)

Hot Spots Winter—Homer Spit in Alaska, Sandia Crest, New Mexico; Mammoth Campground in Yellowstone.
Early Spring—Georgetown, Colorado feeders; Red Rocks Park, Colorado; Road to Guanella Pass, Colorado.
Summer—St. Paul Island in Alaska, Arctic Valley near Anchorage, Alaska; Oregon's Crater Lake National Park.

1

A rather tame bird that breeds in the high and lonely places of Western North America. It nests possibly higher than any other North American bird, surviving and thriving in the harsh and seemingly barren alpine tundra. It is the most widespread of our three species of Rosy-Finches. My lifer Gray-crowneds were not in the high mountains of the West, but a lost group of three that wintered in the garden of the Forestry building on the Fond du Lac Reservation near my home in Carlton County, Minnesota.

Description & ID Tips

A stout brown and pink songbird with gray band rearward from the eyes wrapping around the back of the head, longer wings and notched tail. Females similar to males but show less intense pink on body and reduced black in crown. Three populations can be identified in the field; "Bering Sea" Gray-crowned has gray cheeks, black throat and averages larger than other two. It is found in the Aleutians and other islands in Alaska; "Coastal," sometimes called "Gray-cheeked" or "Hepburn's" have the gray extending down over the cheeks as well as the crown; and "Interior" Gray-crowned with brown cheeks.

Song and Calls

Buzzy chew call with variations including *chew-woo, cheew-wip, chee-up*. Song is not often heard since humans rarely visit its high-mountain summer haunts.

Habitat

Breeding—Above treeline (both altitudinally and longitudinally) in alpine tundra and also on the treeless but boulder-strewn tundra of Alaska's Pribilof and Aleutian Islands. One of the

1. Adult "Interior" Gray-crowned has brown cheeks.

Grand Slam of Rosy-Finches

Listers doing a Big Year (tallying as many bird species as possible during a calendar year north of Mexico) need to be extremely efficient in their travels. So it is no wonder that most make a quick trip to Sandia Crest in the mountains of New Mexico where often all three rosy-finch species can be found feeding at their deck feeders. In a couple hours one can tick off three species that to find on their high mountain breeding grounds would take huge amounts of effort over several states.

Sunflowers gone to seed in a backyard garden provide winter food for a Gray-crowned.

highest altitude nesters in North America.

Winter—Anywhere bare ground and food available: fields, roadsides, wind-swept hillsides, mountain meadows. Deep snows may force them to backyard bird feeders.

Range
Breeding—High western North America including the Rockies, Cascades, Sierra Nevadas and Brooks Range. Also Aleutians and Pribilof Islands of Alaska.

Winter—Western U.S. including Montana, Wyoming, Colorado, Utah, Nevada, California, Washington and Oregon. East to the Black Hills of South Dakota and south to northern New Mexico.

Nest
Ground-nester that builds bulky grasses, lichens, mosses and rootlets wedged Into rocky crevices. Thickly lined with fine grasses and feathers, often from ptarmigan. Typically five eggs, sometimes four or six, rarely three.

2

2. Gray-crowneds forage atop high alpine snow fields in summer for wind-blown seeds and insects.

Food

Grass and weed seeds in winter (along with bird seed at feeders). Adds insects to its diet in summer, and young are fed mostly invertebrates. Does crave salt.

Feeds on and about alpine snowfields at high elevation in summer. Eat cutworms and other insects as they melt out of the margins.

Gular pouches develop in their throat during breeding season allowing both the male and female to transport more food per trip to their developing young.

Nature Notes

Unlike most winter finches, rosy-finches tend to roost in flocks, often using the same rocky outcrop every night. They have even been known to roost in abandoned mud nests of Cliff Swallows.

Winter flocks can number over 1000 individuals and include Horned Larks, Snow Buntings and Lapland Longspurs.

Three Distinct Populations

Three populations can be identified in the field—"Bering Sea" Gray-crowned has gray cheeks, black throat and averages larger than other two. It is found in the Aleutians and other islands in Alaska; "Coastal," sometimes called "Gray-cheeked" or "Hepburn's" have the gray extending down over the cheeks as well as the crown; and "Interior" Gray-crowned with brown cheeks.

There are three subspecies of Gray-crowneds. On the left is the "Hepburns" or "Coastal" that has gray cheeks and "Interior" Gray-crowned with brown cheeks (right).

Brown-capped Rosy-Finch

Description Brown and rosy songbird of high alpine mountains in Colorado and northern New Mexico.

Length 6.25 inches (House-Sparrow-sized)

Other Names *Leucosticte australis* (Scientific), Brown-capped Leucosticte (old name)

Hot Spots Winter—Sandia Crest in New Mexico.
Summer—Trail Ridge Road and Tundra Communities Trail in Rocky Mountain National Park; Summit Lake on the Mount Evans Road in Colorado.

Nearly a Colorado endemic species, the Brown-capped Rosy-Finch also nests in a few isolated mountain ranges in southern Wyoming and northern New Mexico.

Description & ID Tips
Similar to Gray-crowned but with a dark crown. Rosy belly and rose on wings.

Song and Calls
Similar to Gray-crowned Rosy-Finch (see page 33).

Habitat
Breeding—Above timberline high in the southern Rocky Mountains (usually above 12,000 feet). Talus slopes. Highest altitude breeder in North America north of Mexico; one nest found at 14,200 feet!

Winter—Prefer to stay as high as possible in the mountains where food can be found, but deep mountain snows and storms may force them to lower elevations to feed along roads, in brushy fields and at feeders. Always anxious to get back to their high altitude homes as soon as the wind scours bare any patch of mountain tundra.

Range
Breeding—Primarily in the high mountains of Colorado but also southern Wyoming and north-central New Mexico.

Winter—Mountains and basins of Colorado and northern New Mexico.

Nest
Female builds a neat cup nest of dry grasses, stems of flowers

1. Note the Brown-cap's very rosy belly and rose on wings.

Frozen Food

When you nest in high barren alpine scree, you have to get creative in obtaining your daily nutrition, not to mention sustenance for your developing nestlings. Brown-caps find food in, on, and around an unlikely source...melting snowfields.

They scrounge for seeds and insects that have been blown atop the snow patches. Some of these insects are from areas at much lower elevations that have been caught in powerful updrafts during storms. They also glean frozen invertebrates from the thawing margins. Throughout the summer, the melting snow continually reveals edible goodies for the persistant forager.

All rosy-finches feed along the margins and atop summer snowfields at high altitude.

interwoven with moss and lined with fine yellow grasses and feathers of ptarmigan and other finches. Interestingly, the outer rim in some nests laced with the prickly leaves of thistle, presumably as a predator deterrent. She builds quickly (one researcher noted a female making 40 trips in 10 minutes!) and can finish in one to three days. Nest tucked into crevices in boulder talus slopes. Has even been recorded nesting in abandoned mines high in the Colorado mountains.

Food

Snow fields in their high altitude summer breeding range is where you'll find Brown-capped Rosy-Finches searching for wind-blown seeds and insects. They also glean newly-uncovered insect goodies from the melting fringes. Glacier margins provide additional foraging habitat. Seeds in winter but apparently doesn't partake of sunflower seeds like its Black Rosy-Finch cousin.

Nature Notes

Roost communally on winter nights in mine shafts, barns, caves and even the abandoned nests of Cliff Swallows.

Latin genus *leucosticte* translates to "white line," and refers to the high altitude snowfields, their preferred feeding and breeding areas.

Formerly known as the "Brown-capped Leucosticte."

2

2. The author's wife, Bridget, searches for a Brown-capped Rosy-Finch high in Rocky Mountain National Park (see sidebar on next page).

Rocky Mountain High

The search for my lifer Brown-capped Rosy-Finch will always stand out as a highlight of my birding adventures. Not only was the scenery magnificent, but what made it extra special is that I was not alone in this arduous task of huffing and puffing and rock scrambling above 12,000 feet; my girlfriend and future bride Bridget accompanied me. We had a great time searching out the five lifers I hoped to get. She first spotted the Williamson's Sapsucker (without binoculars!), then we ticked off Black Swift and Hammond's Flycatcher. We still laugh about our fifth day of fruitless searching for a White-tailed Ptarmigan when we passed a little boy on the trail who told his mommy about the "little chicken" he'd just seen! (We eventually found a hen with a chick.)

We did finally locate a Brown-cap—and right where it was supposed to be—feeding along the fringes of a high mountain snowfield. John Denver had it right, "Rocky Mountain High, Colorado!"

The author searches for his lifer Brown-capped Rosy-Finch high in Colorado's Rocky Mountain National Park. These high altitude snow fields is where the Brown-caps feed.

Black Rosy-Finch

Description Black and rose sparrow-like finch of high elevations. Only winter storms and deep snow force them to feed at lower elevations.

Length 6.25 inches (House Sparrow sized)

Other Names *Leucosticte atrata* (Scientific), Black Leucosticte (old name)

Hot Spots Winter—Sandia Crest in New Mexico; Mammoth Hot Springs in Yellowstone National Park; Gardiner, Montana.
Early Spring—Georgetown, Colorado feeders; Red Rocks Park, Colorado.
Summer—Top of the gondola (Teton Village) at Wyoming's Jackson Hole Ski Resort on Rendezvous Mountain. For the more ambitious, try hiking the Teton Crest Trail; Mount Washburn in Yellowstone National Park is also a possibility.; Thomas Peak and Wines Peak trails in the Ruby Mountains of Nevada.

L umped with the other rosy-finches from 1983-93, but evidence since then has shown that they are indeed three separate species. I found my lifer near the top of the gondola run at Jackson Hole Ski Resort in Wyoming.

Description & ID Tips
Black where the Gray-crowned is brown. Less rosy on belly than the other two species.

Song and Calls
Similar to Gray-crowned Rosy-Finch (see page 33).

Habitat
Breeding—High, rocky tundra above the timberline (over 11,000 feet) in central Rocky Mountains.

Range
Breeding—Isolated high mountain ranges in Central Idaho, west-central Montana south to southeast Oregon, eastern Nevada, southern Utah and north-central Wyoming.

Nest
Bulky nest hidden deep in a rock crevice, usually in a north-facing vertical cliff in high alpine habitats. Lines nest with hair (Porcupine, Yellow-bellied Marmot) and wool (Bighorn Sheep). Four to five eggs is typical.

Food
Searches high altitude snowfields for wind-blown seeds and insects. Winter diet is seeds—roadside weed seeds, sunflower seeds at feeders and grain in agricultural feedlots.

Nature Notes
Clark's Nutcrackers search out nests and likely predate the eggs and nestlings of rosy-finches.

The High Life

Life above 11,000 feet is not easy, but the Black Rosy-Finch is very well adapted. Returning to their high mountain haunts in April may seem a bit crazy as the snow can still be many feet deep. But they need to be on hand the instant nesting is possible, as the alpine summer is extremely short. Nests are placed in rock crevices overlooking their food source—melting snow fields. Males don't defend *terra firma* territory, but rather a floating circle of space around their mate. As she moves, so does his territory. A pair of sacs below the floor of their mouth allows them to carry extra "edible cargo" back to their sometimes distant nestlings. In this extreme and sterile environment, the extra payload allows them to forage farther afield, sometimes up to 2 1/2 miles from the nest!

Purple Finch

Description Stocky finch of boreal forests that winters over much of the eastern U.S. Males are a beautiful purplish-red; Females streaked and brown with broad white "eyebrow."

Length 6 inches

Other Names *Carpodacus purpureus* (Scientific), *roselin pourpre* (French Canadian)

Hot Spots Winter—Hilton Pond, South Carolina Summer—Widespread in the boreal forest; Northeast Minnesota east to New England.

1

I distinctly remember hearing my first summer singing Purple Finch. It stands out mainly because I didn't know what the heck was warbling from the tip top of a tall White Spruce in northern Minnesota's Boundary Waters Canoe Area. I stood for a long time, scanning for the vociferous songster. I was pleasantly surprised when I finally laid my eyes via binoculars on a male Purple Finch.

Description & ID Tips

Males are not a bright red like House Finches nor are they, by any stretch of the imagination, purple. Purplish-red, maybe. Roger Tory Peterson described them as "a sparrow dipped in raspberry juice." I'll second that. Interestingly, males do not acquire their full red plumage until late in their second summer (see *Why so few Males?* sidebar). Male's breast is deep raspberry red, darker than the rosy red breast of the Cassin's. Large headed with a slight peak (Cassin's can have almost a crested appearance and House Finches have a smaller flatter head).

When trying to differentiate Purples from House Finches, it's best to concentrate on the gals (see sidebar this page).

Song and Calls

A lively and lengthy warble, often from the tip-top of a conifer or other tree. Call note is a sharp, almost metallic *tink*. Song can be mistaken for a Pine Grosbeak when they overlap. Territory song is a rich, warbling and, might I say, cheery song that may be sung in a breathless performance lasting ten seconds or more.

Habitat

Breeding— Habitat generalist that can be found nearly every-

She's Easy

When trying to separate House Finch from Purple Finch, it's always easiest to turn your attention to the ladies. Females are much easier to differentiate. Note the female Purple Finch's broad white "eyebrow," which is visible even at great distance and under poor lighting conditions. House Finch females lack any bold markings and their head is mostly unpatterned.

Note the female Purple Finch's broad white eyebrow.

1. "Dipped in raspberry juice" is how Roger Tory Peterson once described the male Purple Finch.

2

where there are small trees and shrubs and openings. Coniferous forests in the East. Also shrubby woods, parks, open woods.

Winter—Anywhere there is plentiful food.

Range

Breeding—Absent from Alaska but central Canada across southern Canada to the Maritimes. South to northern Illinois, Ohio, Pennsylvania, New Jersey and in the mountains to West Virginia.

Winter—Irruptive. Minnesota and New England south to Florida and east Texas. In the West from southern British Columbia south to Baja California.

Nest

Tree-nester that builds a well concealed cup nest (5 to 60 feet up) that is rarely visible from the ground. Prefers conifers including small ornamental spruces and plantation Christmas trees, but will nest in deciduous trees. Cup made of fine twigs, rootlets and grasses that is lined with wool, moss or hair. Four to five eggs, sometimes three or six.

Food

Poplar catkins and buds. Wide variety of fruits and their seeds including sumac, juneberry, mountain-ash, crabapples. Sunflower seeds at feeders.

2. Purple Finches will begin singing their warbling song in late winter.

3

Nature Notes

New Hampshire's state bird.

Male puts on a show to impress the female during courtship. He'll flutter his wings, hop around, stick out his breast and cock his tail. He'll then launch himself vertically and land, point his beak skyward and lean back as far as he can. All this while singing and holding a piece of nesting material in his beak! Impressive.

3. Female (or juvenile male) Purple Finch feasting on crabapples in a city park.

Why so few Males?

Ever wonder why it seems like there are so few male Purple Finches at your winter feeder and so many dull-colored females? The answer is a bit surprising. A bunch of those "she's" may be "he's." Male Purple Finches don't fully acquire their raspberry red color until late in their second summer, so brown Purple Finches may be either adult females, juvenile females or juvenile males! A nine-year banding study at Hilton Center Pond in South Carolina found that of 2,702 Purple Finches banded, 749 (or 27.7 percent) were red adult males. The other 1,953 brown birds could not be accurately aged or sexed, but if we assume both sexes and all ages winter in the same spots, then about 25 percent were adult females, 25 percent young females and 25 percent young males.

Adult male Purple Finches like this bird are easy to identify. But brown Purple Finches are not necessarily females. They could be juvenile males.

Cassin's Finch

Description Medium-sized finch of western mountain pine forests. Female is streaky brown and male has bright purple-red crown.

Length 6.25 inches

Other Names *Carpodacus cassinii* (Scientific)

Hot Spots Widespread in the West.
Early Spring—Golden Gate Canyon State Park in Colorado; Picnic areas and campgrounds between Steamboat Springs and Highway 14 turn-off to Walden, Colorado.
Summer—Moraine Park Campground in Rocky Mountain National Park, Colorado; Yosemite National Park in California.

A warbling songster of mountain pine and fir forests. Similar to Purple Finch but brighter crown and more brown on back of head and nape.

Description & ID Tips

Males are very similar to male Purple Finches but their upper back and hind neck shows much more brown than red. Male's crown color is brighter than the rest of the head, unlike Purple's whose head is more evenly colored. Also, their belly is unstreaked and white, compared to the Purple Finches much more streaked undersides. Chunky bird that can raise its crown feathers and look almost crested. Female resembles the female Purple Finch but she has more streaking on the face and a less defined white eyebrow compared to the Purple. Her belly is also more heavily streaked.

Song and Calls

Warbling song has been described as "rollicking" and "bright." Call note is a liquid *kee-up* or *tidilip* (compare to the male Purple Finch's metallic *tink* call).

1

1. Female (L) and male Cassin's Finch.

Habitat

Breeding—Variety of coniferous forests in the Interior Mountain West. Usually between 3,000 and 9,000 feet elevation. Mature forests of Lodgepole and Ponderosa Pine. Occasional in open sagebrush with scattered junipers.

Winter—Open coniferous forests. Similar to breeding habitat but at lower elevations.

Range

Breeding—Coniferous mountain forests, often dryer Ponderosa Pine stands. Usually found at higher elevations in the Rockies compared to House Finches and Purple Finches, sometimes nearly to timberline. Eastern Cascades and Eastern Sierra Nevada Mountains. British Columbia through the Rocky Mountain states to northern New Mexico and Arizona.

Winter—Most of breeding range and south to southern California, Baja, southern Arizona, New Mexico, west Texas and central Mexico. Some birds may not migrate and remain in their summer range.

Nest

Tree-nester that prefers the "high life," placing its nest 30 to 40 feet up in a conifer and far out on a branch. Loosely knit cup made of fine twigs, rootlets, stems and lichens with a lining of wool, hair, rootlets and bark fiber. Usually four to five eggs, sometimes three to six.

2

2. Three male Cassin's.

Food

Summer food includes buds, berries and seeds of conifers. Tree buds are a favorite, and one study found that Quaking Aspen staminate buds made up to 94 percent of stomach contents during early stages of nesting. May take Ponderosa Pine seeds directly from cones but also from shed seeds on the ground. Occasionally eats sunflower seeds at winter bird feeders.

Nature Notes

Often associates with other finches including Evening Grosbeaks, Pine Siskins and Red Crossbills. In fact, they sometimes mimic the songs/calls of these three species in their own song.

Some evidence that they may be able to get at the seeds of green pine cones thanks to crossbills prying them open.

Preyed upon by Sharp-shinned Hawks, Cooper's Hawks, Northern Shrikes and Northern Pygmy-Owls.

May be increasing in the north and west parts of its range.

House Finch

Description Native finch to arid western habitats. Introduced to New York in 1940 and spread rapidly west. Now a ubiquitous bird of urban, suburban and other settlements.

Length 6 inches

Other Names *Carpodacus mexicanus* (Scientific), *roselin familier* (French Canadian)

Hot Spots Almost any city, town, hamlet or burg in the U.S. Widespread.

A native species to wild desert habitats in the western U.S. that was illegally introduced to New York as "Hollywood finches" in 1940. Pet dealers fearing prosecution, released the birds and they quickly established a population on Long Island. Now common across the U.S.

Description & ID Tips
Female much more drab (think streaky sparrow) but lacks the bold facial striping of the female Purple Finch (see *Description & ID Tips* under Purple Finch on page 43).

Song and Calls
Song a descending warble often ending on a *veeerrr* note.

Habitat
Year-round—Around buildings in urban, suburban and rural settings. In West, undisturbed desert and shrub habitats.

Range
Year-round—Originally a native to the western U.S., now found across the continent from British Columbia across southern Canada south to Florida, Mexico, California.

Nest
Cup of fine grasses, wool, string, etc. placed in tree cavities, cacti, tree branches, nest boxes. Seem to have a special attraction to nesting in hanging flower baskets in urban/suburban areas. Usually four to five eggs, sometimes two to six.

Food
Diet is 97 percent seeds, fruits, leaves, buds and flowers.

Nature Notes
The eye disease mycoplasmal conjunctivitis reduced the eastern U.S. population by 60 percent between 1994-97.

Meeting in the Middle

After the House Finch's clandestine beginnings on the East Coast in 1939-40, they rapidly spread across the eastern U.S. reaching Maine, Wisconsin, Missouri, Alabama and Georgia by 1980. Hard to believe now, but in 1986 I traveled from my home in Duluth to Marshall, Minnesota just to add it to my state list, a round-trip of 550 miles! Today they are often considered a "trash bird" when seen during birding outings. Their range now extends from coast to coast.

House Finches commonly nest in hanging flower baskets. They have also nested in Christmas wreaths and atop light fixtures.

Red Crossbill

Description Wandering resident of pine forests who uses its crossed bill to access pine cone seeds.

Length 6.25 inches (House-Sparrow-sized)

Other Names *Loxia curvirostra* (Scientific), Common Crossbill (Europe), *bec-croise des sapins*, *Le Bec-croise rouge* (French and French Canadian), *Mindre korsnäbb* (Swedish), *Kruisbek* (Dutch), *Fichtenkreuzschnabel* ("Pine Crossbill" in German), *Grankorsnebb* (Norwegian), *Krossnefur* ("Crisscrossbill" in Icelandic), *Pikkukäpylintu* (Finnish)

Hot Spots Winter—Superior National Forest, Minnesota.
Summer—Along the mountain road to Hurricane Ridge in Washington's Olympic National Park; Moraine Park Campground in Rocky Mountain National Park; Yosemite National Park; Utah's Wasatch-Cache National Forest.

1

R ed Crossbills are pine specialists, their bills adapted to feeding on the cones. They rarely come to feeders but will pick salt on rural roads, flying up in front of your car at the last second. "*Jip, jip*" call notes are different than White-winged calls and can be used to identify flocks as they zip overhead.

Description & ID Tips

Males are a duller and darker red than male White-wingeds and lack the white wing bars. Females olive-gray with greenish-yellow head, breast and rump. Large-headed, large-billed, shorter-tailed and only slightly smaller than their White-winged cousins, but this is not noticeable in the field.

Song and Calls

Winter flocks are separated from White-wings by their hard *jip jip jip* calls, but note that there are many variations of flight calls (see *Ten Species in One?* sidebar on page 55). Song is a short series of buzzes and trills interspersed with *jip* notes.

Habitat

Year-round—Pine forests with edge habitat including isolated clusters of trees. In the Rocky Mountains they prefer forests of Ponderosa Pine and Lodgepole Pine. White Pines and especially Red Pines are preferred in the Great Lakes and New England.

Range

Breeding—Circumboreal. Found across coniferous forests from North America and Eurasia. In North America from southeast Alaska to Newfoundland south to the mountains of Baja California and northern Nicaragua. Year-round resident

Nesting in Winter?

Amazingly, this species can nest in any month of the year if they settle into an area with an abundant cone crop. Studies by Bailey et al., 1953 found that they could reproduce at temperatures ranging from 45° F to -9° F. In fact, midwinter is the peak of breeding in Colorado. One biological quirk that makes nesting in a season other than summer possible is that the young are able to digest and thrive on seeds regurgitated by their parents.

Red Crossbills have been recorded nesting in every month of the year!

1. Male Red Crossbill in a pine. They specialize in prying seeds from pine cones.

2

Nest

Tree-nester that usually utilizes coniferous trees (especially pines) to place its well-concealed cup nest. Nest is usually placed on a limb well out from the trunk. Has used Engelmann Spruce and Subalpine Firs in the Rocky Mountains. Twigs, leaves, moss and grass form the nest structure that is then lined with wool, hair, feathers, moss and lichens. Despite their broad range, very few nests have ever been found. Usually three to four eggs, rarely up to five.

Food

Seeds of conifers, especially from pine cones. They open cones by inserting their closed bill between the cone scales and then pry scales apart by opening their crossed mandibles. Seeds are extracted with their nimble tongue and quickly and efficiently husked. Like their white-winged cousins, they only rarely show up at backyard bird feeders in winter.

Nature Notes

Ornithologists are not in agreement on what to do with Red Crossbill systematics; Some prefer calling these types subspecies, others would prefer the term "pseudospecies" and yet others want to create TEN new species, which would be a nightmare for the birder trying to separate them in the field.

Calcium seems to be a craving for Red Crossbills. Observers have noted them picking at calcium chloride spread on a dirt road, calcium from cliff walls, putty around windows, old deer bones, bone fragments in carnivore scat, charcoal and wood ashes.

from the southern boreal forests of Canada south to New York and New England, Minnesota, Wisconsin and northern Michigan, mountains of California, Rocky Mountains, and even into the high pine-covered peaks of Mexico, Guatemala, Honduras and even Nicaragua. In Eurasia, Scandinavia, British Isles, mainland Europe (Alps, Pyrenees) and across Asia (taiga and mountains) to Siberia and Japan. Other forms found in Himalayas, Vietnam, Philippines and Atlas Mountains of Morocco in north Africa.

Winter— Anywhere there is a good pine cone crop. In the Rockies this includes stands of Pinyon Pine.

2. Red Crossbill flock feeding on Tamarack cones.

Ten Species in One?

Ten North American types north of Mexico have been identified based on distinct flight calls, which may reproductively isolate each group from others (though this has not yet been proven). In studies, flocks tended not to respond to different flight calls than that of their own type. Are they ten distinct species? That is the million dollar question. Each type has a bill that is shaped and sized slightly differently, an adaptation to opening the cones of different species of conifers, including pines (*Pinus*), spruces (*Picea*), Douglas-firs (*Pseudotsuga*), hemlocks (*Tsuga*) and tamaracks/larches (*Larix*). Complicating the identification mess is the fact that several groups may migrate together and even breed near one another. The one nicknamed "Old Northeastern" race (Type 1), nearly became extinct in the late 1800s as the New England White Pine forests were cleared and the Appalachians were logged off.

Type 1 ("Appalachian Crossbill")—Red Spruce, White Pine; Core range: Appalachians, irruptive in Eastern U.S.; bill medium, flight call *kyep,* descending

Type 2 ("Ponderosa Pine Crossbill")—Ponderosa Pine; Core range: Continent wide and irruptive; second largest bill; flight call *kewp,* descending, deeper than Type 1

Type 3 ("Western Hemlock Crossbill")—Western Hemlock; Core range: Pacific Northwest, irruptive to Great Lakes; smallest bill; flight call *chik,* squeaky, little inflection

Type 4 ("Douglas Fir Crossbill")—Douglas Fir; Core range: Pacific Northwest; bill medium; flight call *kwit,* bouncy and rising, less squeaky than Type 10

Type 5 ("Lodgepole Pine Crossbill")—Lodgepole Pine; Core range: Interior West; bill medium-large; flight call *clip,* "twangy" higher and weaker than Type 1

Type 6 ("Sierra Madre Crossbill")—Apache Pine, etc.; Core range: Mexico, wanders to southwest U.S.; largest bill; flight call *teew,* clear, nearly whistled

Type 7 ("Enigmatic Crossbill")—Western Larch?, Western White Pine?; Core range: Pacific Northwest; bill medium; flight call *chek,* with little inflection

Type 8 ("Newfoundland Crossbill")—Black Spruce; Core range: Newfoundland only; bill medium-large; flight call *chilp,* "ringing, complex"

Type 9 ("South Hills Crossbill")—Lodgepole Pine; Core range: Southern Idaho; bill medium-large; flight call *pik,* low and hard

Type 10 ("Sitka Spruce Crossbill")—Sitka Spruce; Core range: Oregon coast to northern California; bill medium-small; flight call *pweet,* rising strongly and squeaky

*Adapted from The Sibley Guide to Birds: Second Edition (2014) and eBird.org online article North American Red Crossbill Types: Status and Flight Call Identification (8 Oct. 2012)

White-winged Crossbill

Description Elusive wanderer of boreal forests that uses its crossed mandibles to pry open scales of Black Spruce and Tamarack cones.

Length 6.5 inches

Other Names *Loxia leucoptera* (Scientific), Two-barred Crossbill (Europe), *bec-croise bifascie* (French Canadian), *Bändelkorsnäbb* (Swedish), *Witbandkruisbek* (Dutch), *Bindenkreuzschnabel* ("Banded Crossbill" in German), *Båndkorsnebb* (Norwegian), *Víxlnefur* (Icelandic), *Kirjosiipikäpylintu* (Finnish)

Hot Spots Winter—Sax-Zim Bog in northeast Minnesota; St. John's, Newfoundland. Summer—Alaska's Denali National Park and environs.

It pays to learn the "*wink, wink*" call of the White-winged Crossbill since they rarely come to feeders. Your first encounter is more likely to be a brief view as they fly overhead or feed in spruce tops in boreal forests. Either way, we are almost always straining our necks as we look up at this remarkable bird. Crossed mandibles pry apart spruce cone scales to get at the seeds. Larger than redpolls yet smaller than winter grosbeaks, their namesake paired white wingbars cinch their identification. Males seem a brighter red than their male Red Crossbill cousins. White-wingeds show a strong preference to feeding on spruce and tamarack cones, while the larger billed Reds primarily feed on the cones of pines. Sometimes found on plowed roads where they forage for salt (see sidebar on page 59).

Description & ID Tips

Both sexes have two white wing bars (more bold on adults). Longer tailed than Red Crossbill and with a smaller rounder head and smaller bill. Female's undersides are olive-yellow, male's pinkish. In summer, male's belly more red. Juveniles and females have brown flight feathers while adult male's are black.

Song and Calls

Flight call is a sharp *wink wink…wink wink wink*. They also give a rapid series of *chet chet chet* calls. Song is a long series of dry staccato trills, each phrase at a different pitch.

Habitat

Breeding—Boreal forest and peatlands. Anywhere large stands of spruce and tamarack have produced bumper seed cone crops.

Parrot of the North

To brand-new birdwatchers who do not yet know the name of this bird, its bill must resemble a severe and crippling deformity. Surely no bird can eat with a bill like that! But their bill is highly specialized for feeding on cones of conifers. Their closed bill is inserted in a cone then opened to pry apart the scales to get at the seeds. A flexible tongue with a spoon-shaped tip scoops up the freed seed. They then go to work husking the seed before eating it.

White-wings earn their nickname, "parrots of the north" by clinging sideways and even upside down to the cones of spruce as they feed.

1. Male White-winged Crossbill clings to a White Spruce cone as it feeds.

Winter—May wander south of the boreal forest during irruptions and feed in stands of large spruces in towns, cities, parks, farms and even rural cemeteries on the prairie.

Range
Breeding—Circumboreal. Looking at the distribution map for this nomadic wanderer one might get the impression that they are quite common and widespread, but the fact is they are so dependant on good spruce cone crops that they may be absent from large areas of their range in any given year. We also chose to combine their wintering and breeding ranges as they can nest even in midwinter! Dashed line indicates southern limit of winter wanderings.

Boreal forest habitat must have spruces and/or tamaracks. Can be found from central Alaska across Canada and south to New England, Minnesota,

Wisconsin and northern Michigan; Cascades of Washington and Oregon; and the Rocky Mountains south to Wyoming, Colorado and New Mexico.

Migration—Up to 10,000 have been recorded moving through an area of Ontario in a single day.

Winter—During irruptions they may wander far south of the boreal forest to the central U.S. There are even rare records from Florida and Texas.

Nest
Tree-nester with nest placed at varying heights (6 to 60 feet) in a spruce. More often than not high up and far out on a branch. Spruce and tamarack twigs, leaves, moss, birch bark and grass form the nest structure that is then lined with wool, hair, feathers, cocoons, moss and lichens. Usually three to four eggs, rarely up to five.

2. Female White-winged Crossbills are yellowish-olive.

3. White-wings are more rosy-red than the brick-red Red Crossbill.

Cravings: Otter Poop & Road Salt

Coprophagia is not a word you hear too often in the bird world, or for that matter, ever hear at all. It is the act of feeding on feces (i.e. eating poop). Much more common in the insect and mammal realm, it is nearly unheard of among birds. But while doing otter research in Kouchibouguac National Park in New Brunswick, Canada, Daniel Gallant observed two White-winged Crossbills fly down and land on fresh scat of a River Otter and proceed to eat chunks. Possibly they were taking in calcium from the bones in the scat.

Crossbills seem to crave salt though it is not thought to be an essential element to their diet. Anyone who has birded the North Woods in winter has seen small flocks of crossbills scatter from the road ahead while feeding on road salt. But research on Red Crossbills suggests that the birds get plenty of sodium in their normal diet. They have been noted feeding on a bizarre list of items including, mud in roadside puddles, dog urine on the snow, salt lick blocks put out for deer, and even leftovers from salt pork barrels (This was in 1888 when there was such a thing as a salt pork barrel).

Flocks of White-winged Crossbills often pick salt off rural roads in winter.

A Tale of Two Bills

The common birder wisdom is that White-winged Crossbills are more often found foraging on spruce and tamarack cones, while Red Crossbills seek out pines with good cone crops. But Craig Benkman of State University of New York wanted to prove this. He found that White-winged Crossbills with their more slender and shorter bills were able to extract and husk seeds from smaller cones quite a bit faster than Red Crossbills. Conversely, Red Crossbills were very efficient extracting seeds from larger cones, husking Red Pine seeds twice as fast as White-wings and more than three times faster than White-wings at White Pine, the largest-coned pine in northeastern North America. White-winged Crossbills key in on open Black Spruce cones and are very efficient at feeding on them. Benkman estimated that they need to remove, husk and consume from 2,600 to 3,160 seeds in a day! Their small bill is not efficient at opening closed Black Spruce cones. The larger bill of the Red Crossbill is able to open the closed Black Spruce cones but in mixed forests they much prefer the cones of pines.

White-winged Crossbills not only prefer smaller coned trees such as this White Spruce, but are much more efficient at accessing the seeds than Red Crossbills.

Food

Almost exclusively the seeds of spruce and tamarack, but occasionally fir and pine seeds, and even more rarely seeds of deciduous trees, grasses, tree buds, spiders and insects.

Sporadic and spotty distribution of spruce and tamarack cone crops make this species one of the most erratic wanderers of all the winter finches. Rarely shows up at backyard bird feeders in winter.

Nature Notes

Races of this same species occur across northern Europe from Norway east to Siberia. The Hispaniola Crossbill (*Loxia megaplaga*) which lives in the high pine forests of Haiti and the Dominican Republic was formerly considered a subspecies of the White-winged Crossbill.

Crossbill bills can cross to the left or right on different birds.

Amazingly, White-wings may nest in any month of the year! Whenever a pair intuitively determines that the local cone crop is sufficient (for egg development and to last through three weeks of the nestling stage), they will go for it. Winter nest usually placed on the south to east side of the tree to take advantage of solar warmth. Nests built in winter probably larger than summer nests, presumably for more insulative value.

Studies in the early 1960s by Minnesota's Harrison "Bud" Tordoff and William R. Dawson found that crossbills maintained elevated metabolic rates overnight in winter. How do they do it? Good feather insulation for their size helps, as does roosting in dense conifer foliage at night. Large crops enable them to store and digest seeds at night helping to generate heat. They also feed in flocks, and will signal other groups flying by that they have found a bonanza of food (sidebar right).

Rapid Assessment Team

Crossbill flocks are a "rapid assessment team" when they begin foraging in a stand of spruce or tamarack. The flock notices individuals who seem to be finding seed-laden, ripe and open cones and they join in the feast. Since individual trees can vary in the number of seeds per cone and ripeness, this is an efficient way of getting the best "bang for the buck" in caloric intake. Crossbills are also able to instantly determine if a seed husk is empty and they let it drop to the ground, but less than 1 in 1,000 good seeds is dropped accidentally.

One way to locate crossbills is to watch for husks and scales drifting down from the top of a large pine or spruce.

Common Redpoll

Description Gregarious small finch of the Far North that often irrupts to the "south" of the northern U.S. and southern Canada.

Length 5.25 inches (warbler-sized)

Other Names *Carduelis flammea* (Scientific), Redpoll (Europe), *sizerin flamme* (French Canadian), *Gråsiska* (Swedish), *Barmsijs* (Dutch), *Birkenzeisig* ("Birch Siskin" in German), *Gråsisik* (Norwegian), *Auðnutittlingur* (Icelandic), *Urpiainen* (Finnish)

Hot Spots Winter—Ontario's Algonquin Provincial Park Visitor Center, Minnesota's Sax-Zim Bog.
Summer—Churchill, Manitoba.

1

A gregarious winter visitor whose lively chatter and jaunty red "beret" makes them a winter feeder favorite. But they are an irruptive species that only makes forays into the Lower 48 during years of poor birch and alder seed crops in Canada and Alaska. The winter of 2013-14 found ZERO redpolls wintering in northern Minnesota, while the following year, 2014-15, saw a massive influx of the red-capped fluffballs. It is hard to estimate the actual numbers that invaded the North Woods that winter but likely hundreds of thousands. Almost anywhere one pulled the car over in wooded or swampy country and pished, a flock would materialize.

Description & ID Tips
Looks smaller than a chickadee, mainly because the redpoll is a bit slimmer than the fluffy ball-like chickadee (especially on below zero days). Red cap ("poll") and tiny size is diagnostic. Male has a red breast and flank. Brown streaking on sides is heavier than on a Hoary. Bill is sharper and longer than Hoary's, but not as long and pointy as Pine Siskin's.

Song and Calls
No real territorial song but their chattering increases in spring. Call is a lively series of short calls (*che…che…che*) and trills. In flight they occasionally add a burry, rising whistle reminiscent of the Pine Siskin's but not as long—*meeee? meeee?*

Habitat
Breeding—Taiga and boreal forests of stunted spruce, birch scrub and mixed conifer-birch woods.

1. Male Common Redpolls show rosy red breasts.

Survival at Minus 30

Since most of our winter finches spend the cold season either in Canada, the northern tier of states or at high elevation, they end up dealing with bitterly cold conditions. One survival strategy is piloerection, which means extending their feathers out from their body. Small muscles beneath the skin contract when the bird is cold, and this causes the feathers to stand up from the skin. This creates more air space around the feathers and provides more insulation against the cold. Mammals have a similar mechanism for their fur; human mammals get goose bumps!

Piloerection is not a dirty word! It is the strategy for staying warm by fluffing your feathers.

Alder seeds are a favorite winter food of redpolls. Here two Commons forage at dawn in a Speckled Alder. They are very acrobatic, clinging to tiny twigs, cones and catkins, sometimes even upside down, to get at the seeds.

Winter—Northern forests with good seed crops of birch and alder, and backyard bird feeders.

Range

Breeding—Circumboreal. A resident of far northern boreal forests to subarctic lands across Siberia, northern Scandinavia to British Isles and Iceland. Breeds across northern Canada and northern Alaska north to Baffin Island and coastal Greenland.

Winter—Irruptive. In years of abundant food redpolls may stay in their far north breeding range.

Seems to be a NW-SE axis of migration from Alaska through the Prairie Provinces to New England usually invading the northern U.S. every few years. Found in areas with good seed crops of birch and alder. Most winters at least a few can be found in the northern tier of the United States but during massive irruptions may wander as far south as northern California, far north Texas, Arkansas and North Carolina.

Nest

Tree- and shrub-nester that builds an "untidy" nest of plant stems and twigs, three to six feet off the ground in a small spruce, tamarack or willow (17 of 19 nests in northern Ontario built in Arctic Willow (*Salix arctica*)). May rarely nest high in a tree. Lined with feathers (Spruce Grouse, ptarmigan), hair and downy seeds of plants. Often found in loose associations of nesting pairs. In the Arctic, where presumably nesting sites may be limited, they will reline and re-use and old nest. Four to five eggs, sometimes three to seven.

2. Most of us will probably never see a redpoll nest. Their breeding range is in the sparsely-populated Far North.

"Yellowpoll"

In birds, both red and yellow feathers are pigmented with carotenoids (think carrot). The rare condition where yellow replaces red is called xanthochroism. It is genetic and has been recorded in Common Redpolls, Northern Cardinals, Rose-breasted Grosbeaks, Scarlet Tanagers, Red-bellied Woodpeckers, Evening Grosbeaks and several warblers.

Yellow pigment has replaced the normal red-pigmented cap feathers of this xanthochroistic redpoll. It visited the feeders at the Sax-Zim Bog Welcome Center in the winter of 2012-13.

Food

Eats seeds of birch, basswood, alder, willow, elm, spruce, tamarack, grasses, weeds, and of course, our backyard "thistle" (niger seed) and sunflower seed from feeders. Acrobatic forager, often clinging upside down to catkin or small twig in order to get at seeds. Has been noted in Wisconsin shaking birch catkins free of seeds then dropping down to feed on the fallen seeds atop the snow. Birch seeds are an especially high energy winter food. Wildly fluctuating winter numbers are thought to be associated with food supplies farther north and especially tied to the two-year birch seed production cycle. In summer will prey on insects, primarily to feed their young.

Winter Fuel & Winter Warmth

A bird that winters in the "frozen tundra" of Canada, New England and the northern Great Lakes has got to have special adaptations to life in the cold. And redpolls do. Firstly, they go into winter with double the number of feathers they had in summer. Alaskan redpolls had 31 percent more feathers by weight in November than July. They are also masters at fluffing their feathers ("piloerection") and can reduce heat loss even more by inactivity, roosting in dense foliage and tucking their head into shoulder feathers (see photo bottom right).

Conversely, redpolls do not deal well with heat. Studies have shown that they become lethargic and very thirsty at high temperatures. 100°F is highest temp they can tolerate.

An abundant source of seeds is critical to winter redpoll survival. Seeds of birch and alder are especially important, as is the niger ("thistle") and black oil sunflower seeds we offer at our feeders. I have also seen them extracting seeds from spruce cones. Expandable pouches off the redpoll's esophagus called diverticula can be packed with seeds quickly in late afternoon. Then while sheltered in a dense conifer for the night, the redpoll can regurgitate its throat-cache and husk and swallow the seeds. This useful habit can raise their metabolism and save much energy on extremely cold nights.

Scan to see video of redpolls in the wild and at a feeder.

Niger seed at feeding stations is a welcome food source (left), but many forage on wild foods too, like this bird picking seeds from an alder cone (middle). This redpoll's head is turned 180° and tucked deep into its shoulder feathers to retain warmth (right).

Dangerous World

There's not a lot of meat on a redpoll but there must be enough caloric reward for Peregrine Falcons, Merlins, and jaegers, all who take adults during the breeding season. Nest predators include Gray Jays and Red Squirrels. This winter in northern Minnesota I heard of two eyewitness accounts of a Northern Hawk Owl taking a redpoll (see photo below left). This has also been recorded in the literature in other areas. Obviously Northern Shrikes staking out a backyard bird feeder will not hesitate to take a redpoll as one did at my home feeder in 2009 (photo below right). Flocks can provide safety in numbers via early detection of predators, but when preoccupied with feeding, sometimes it's not soon enough.

Avian predators of the Common Redpoll include the Northern Hawk Owl [Superior National Forest, Minnesota] and Northern Shrike [atop redpoll at author's bird feeder in Carlton County, Minnesota].

Nature Notes

In 1867 someone in England had the crazy notion to introduce redpolls to New Zealand. From that initial release of 14 birds, the redpoll has now colonized the "Land of the Kiwis" and nearby islands.

One redpoll banded in Michigan somehow found its way to Okhotsk, Russia. One banded in Belgium was recovered two years later and 5,185 miles away in Heilongjiang Province of China!

Recent genetic studies suggest that Common and Hoary Redpolls may actually be one species. This would make some birders very happy and upset many others (including this author!).

"Goggles" & "Patches": Leucistic Redpolls

Often mistakenly referred to as "partial albinos," most redpolls with aberrant white plumage are actually *leucistic*, a word that comes from the Greek for white (*leukos*). It is a fairly common condition in the Common Redpolls, and any irruption winter will likely turn up a few such birds. The ones shown below were nicknamed "Goggles" (at my feeder in Carlton County, Minnesota in 2007), and "Patches," who was a regular visitor to Minnesota's Sax-Zim Bog Welcome Center in the winter of 2014-15. They often become mini-celebrities at public feeders as watchers always enjoy being able to identify an individual bird from a big flock.

Leucism is a condition where the affected feathers lack pigment. It is possible to have a completely white leucistic bird, but its bill and eyes will be colored normally. Albinos will always have pink eyes.

Not partial albinos! These redpolls show some leucistic feathering.

Greater Redpoll

The subspecies of Common Redpoll that breeds on Canada's Baffin Island and Greenland is known as the "Greater" Redpoll (*Carduelis flammea rostrata*). They occasionally show up in southern Canada and the Lower 48 in winter. Larger than "southern" redpolls, with a darker face, and throat, wider and darker streaking on flanks. Can appear to be shaped more like a House Finch than a redpoll.

"Greater" Redpoll is a subspecies of the Common that is larger and darker [Sax-Zim Bog, Minnesota (top) and Iceland (bottom)]

3. Male (right) and female Common Redpoll.

4. Trusting redpoll eats out of the author's hand.

How to tell a Hoary from a Common

1. OVERALL APPEARANCE & BACK
Hoary—Frosty white; Male has pinkish wash on breast
Common—Brownish; Male breast shows bright red

2. STREAKING ON FLANKS
Hoary—Few, thin and not well defined; Grayish
Common—More, thicker and darker; Brownish

3. RUMP
Hoary—Unstreaked and pure white
Common—Streaky and brownish

4. UNDERTAIL COVERTS
Hoary—White with few streaks (only 1 on bird below)
Common—Three or more streaks

5. BILL & FACE
Hoary—Bill small and conical; Face appears flatter with a steeper forehead; Face has "pushed in" appearance
Common—Bill longer and pointier; Forehead slopes

**Note that not all redpolls can be identified in the field. Some are just best left as "redpoll species"

Common

Hoary

Hoary Redpoll

Description Rare frosty-white small finch of the Far North that may be seen with flocks of Common Redpolls in southern Canada and extreme northern Lower 48.

Length 5.5 inches (warbler-sized)

Other Names *Carduelis hornemanni* (Scientific), Arctic Redpoll (Europe), *sizerin blanchatre*, Snösiska (Swedish), *Witstuitbarmsijs* (Dutch), *Polarbirkenzeisig* ("Polar Birch Siskin" in German), *Polarsisik* (Norwegian), *Hrímtittlingur* (Icelandic), *Tundraurpiainen* (Finnish)

Hot Spots Winter—Minnesota's Sax-Zim Bog in winters with a major redpoll irruption; Algonquin Provincial Park in Ontario; Feeders in Winnipeg, Edmonton, Saskatoon, Canada.
Summer—Churchill, Manitoba.

1

Hoaries are a favorite of mine. Just the simple fact that they have flown hundreds to a couple thousand miles from their Arctic home to my home is amazing to me. They are much rarer than their "Common" cousin. Minnesota bird guide, Kim Eckert, believes the ratio is about 1 Hoary per 100 Commons. I also love their frosty white plumage and trusting nature; they often allow close approach when busy gorging on alder catkins along rural roads.

Description & ID Tips

First I want to note that there is no shame in not being able to separate a light Common Redpoll from a dark female Hoary, nor in saying, "I guess that's just a Common." When trying to identify your first Hoary, start with a very overall white bird that catches your eye. Is the back frosty white with no brown? Does the bill look stubby and conical and the faced "pushed in?" (due to dense feathering around the bill.) Is the streaking on its sides subtle and fine? Is the red on the cap ("poll") and the black around the bill more limited than in a Common? If it's a male, is the color on the breast pinkish and pale? (sidebar photo on this page) Now, to cinch the deal, is the rump unstreaked? Simpy, dimpy! Of course, we all hope for that magical sighting of a "fluffball" Hoary that looks like a snowball in a shrub. These Hoaries can be identified even while driving 50 mph down a road. Of course, it's always best to stop and get a really good look….and a photo.

Song and Calls

Lively series of short calls and trills. In flight they occasionally add a buzzy trill reminiscent of that of the Pine Siskin's but not as long nor rising.

1. A classic frosty white Hoary Redpoll [Sax-Zim Bog, Minnesota].

Hoary:Common Ratio

On southern Manitoba Christmas Bird Counts, Hoaries rarely exceed 2 percent of all redpolls. Breeding populations in Churchill, Manitoba fluctuate wildly with Hoaries constituting from 5 to 50 percent of the total nesting redpolls. In northern Minnesota, bird guide Kim Eckert feels that 1:100 is the likely ratio of Hoaries to Commons seen during winter irruptions. Interestingly, Leroy and Myrtle Simmons, who banded over 22,000 redpolls in their Winnipeg backyard, and obviously were able to observe them in the hand, tallied 3.7 percent of all redpolls as Hoaries. Maybe birders are a bit conservative.

Male Hoaries show a pinkish wash on their breast, unlike Common male's bold red.

Habitat

Breeding—Open taiga forest habitat broadly overlaps with Common Redpoll, but Hoaries will nest farther out onto the tundra, as far as shrubs extend.

Winter—Northern forests with good birch and alder seed crops. Bird feeders in northern towns.

Range

Breeding—Circumboreal. Breeds slightly farther north than the Common Redpoll, mostly at latitudes above 60 degrees N including Ellesmere Island and Greenland, but dips south along the Hudson Bay coastal tundra to the Churchill, Manitoba area. In Eurasia, from Norway across Scandinavia to Siberia.

Winter—Not often seen in the Lower 48, but can be fairly common some years in northern Minnesota and southern Canada. Some stay in the Far North and how they survive in the perpetual darkness is still a mystery.

Nest

Shrub- or ground-nester that makes a cup nest of grasses, twigs and rootlets. Nests placed one to seven feet up in shrub, but may also call a rock-sheltered ground-nest home. Lined with plant down, feathers and hair. Will reuse old nests after giving them a fresh lining. Four to five eggs, rarely three or six.

Food

Similar to Common Redpolls; seeds of trees (birch, spruce), shrubs (alder, willow), weeds and grasses.

2 & 3. Hoary feeding on Tansy (top) and Speckled Alder seeds (bottom).

4

Nature Notes

Studies of captive Alaskan redpolls showed that Hoaries can survive temperatures as low as -89°F while Commons do not survive below -65°F. How do Hoaries survive at these brutal temperatures? Hoaries have more insulating plumage than Commons, continued activity at lower light levels than Commons, and were able to gain weight and maintain fat at temperatures below 23°F.

"Barrel-chested" Greenland Hoaries

The Greenland subspecies (*C. h. hornemanni*) breeds in the High Arctic of northern Greenland and adjacent Canada. It is generally paler and definitely larger than *C. h. exilipes*. They are longer-tailed and thicker-necked and may appear barrel-chested and "front heavy." But be aware that you really need to have other redpolls of other subspecies nearby for comparison to make this call. *C. h. exilipes* is the low-arctic Hoary that is most often seen in southern Canada and northern U.S.

"Greenland" Hoaries photographed in Iceland. Note bottom bird's large size and "barrel chest."

4. A "fluffball" Hoary gets a "drink" of snow after feeding on alder seeds.

Pine Siskin

Description Irruptive finch of boreal and mountain forests. Yellow-tinged wings and sharp pointy bill. Gregarious in winter and often aggressive at bird feeders.

Length 5 inches (warbler-sized)

Other Names *Carduelis pinus* (Scientific), *tarin des pins* (French Canadian)

Hot Spots Widespread.

1

A feisty winter visitor to our bird feeders that irrupts in massive numbers in years of poor seed crops in its breeding range in Canada.

Description & ID Tips

Three things catch my eye when identifying a siskin. In order of importance: thin and sharp bill; yellow wing bars; deeply forked tail. These traits may seem obvious when one lands on your feeder, but when high up in a roadside birch tree, the field marks are essential to separate them from silhouetted redpolls and goldfinches.

Song and Calls

Most distinctive is their buzzy ascending trill—*zhreeeeeeee!* I often listen for that call when a flock flies overhead in winter. It is a sure way to tell them from a flock of redpolls. Song is a jabbery rapid series of calls, frequent twittering and chattering like redpolls, but with a distinctive rising *zhreeee*. Some phrases are reminiscent of House Sparrow's call and some sound like a softer trill of an Evening Grosbeak.

Habitat

Coniferous forests.

Range

Breeding—Widespread but they are actually localized and erratic breeders within the broad range. Alaska, Yukon and Northwest Territories, south to Mexico in the west, and east to the Canadian Maritimes, New England, northern Michigan, Wisconsin and Minnesota.

Winter—Irruptive. Wanders erratically and widely over the U.S. Some years south to Florida, Texas and Mexico.

The Great Siskin Invasion

Cornell Laboratory of Ornithology's Project Feeder Watch estimated a minimum of 95 million Pine Siskins at feeders in the winter of 1987-88. They were found in nearly all corners of the country. The following year they seemed to vanish, likely finding plenty of food in the sparsely populated core of their breeding range in Canada. I vividly remember the cacophony of calling siskins as flocks amassed around my home in Duluth, Minnesota that winter.

Massive irruptions of Pine Siskins occur occasionally in the U.S.

1. Look for the Pine Siskin's yellow "T" on each wing. Some individuals are more yellow than others.

2

Nest

Tree-nester that often places its nest on an outer branch in a well-concealed fork of a spruce. May be low to high in the tree. Loose congregations of nests are not unheard of. Nest is relatively large compared to the size of the bird, and made of grass, twigs and rootlets, lined with fur, fibers, feathers, rootlets and hair. Three to four eggs usually, sometimes three or five or even six.

2. A swirl of Pine Siskins erupts from a Yellowstone meadow in late fall.

Food

Seeds of birch, spruce, cedar, fir and weeds. Observed staking out territory next to much larger Evening Grosbeaks who in the process of cracking open sunflower husks would drop bits of the seed. The lurking siskins would quickly snatch the bits.

Nature Notes

Manitoba overwintering population increased dramatically from the early 80s to the mid 90s, then decreased.

3

4

3. My son Bjorn hand feeds a trusting Pine Siskin.

4. Birch seeds littering the snow's surface is a sign of feeding siskins.

Aggression

Feisty little buggers at feeders, siskins constantly bicker amongst their own kind and will even chase off redpolls, chickadees. Aggression display is shown in photo below—lunging with head lowered, wings and tail spread. They even utter a threat call. If this fails, they will sometimes continue the battle in the air, combatants occasionally rising to five or ten feet while still embroiled in the conflict.

Siskin battles at winter food sources are common, and good entertainment for feeder watchers.

American Goldfinch

Description Much-loved finch that many of us can enjoy in winter and summer. Loses its bright black and yellow plumage in late fall.

Length 5 inches (warbler-sized)

Other Names *Carduelis tristis* (Scientific), *chardon-neret jaune* (French Canadian)

Hot Spots Widespread and very common.

1

Though the much-loved goldfinch loses its bright yellow and black plumage in winter, it's still a favorite at many a winter bird feeder. Its cheerful *chip dip….potato chip* call is diagnostic and helps separate it from other small finches flying overhead.

Description & ID Tips

Don't look for the yellow and black plumage that we all know and love at your winter feeding station. Both sexes lose their bright breeding plumage and become drab winter finches. A glance at their unstreaked and evenly-colored body quickly separates them from redpolls and siskins. Next note their buffy wingbars and pale pinkish bill. By March, the males begin to show brighter yellow plumage.

Song and Calls

Summer song is a lively, loud series of sweet and loud repeated phrases. Can be confused with Indigo Bunting so listen carefully to separate them at a distance. Also listen for the *chip chip potato chip* flight call.

Habitat

Year-round—Habitat generalist that can be found nearly everywhere there are small trees and shrubs with openings.

Range

Breeding—Winters from Washington to Minnesota to Maine and south to Florida and Mexico. Continent wide and south to California, Colorado, South Carolina.

Winter—Population shifts south in late fall but numbers vary from winter to winter in the northern part of range.

1. Winter plumage of the American Goldfinch.

The Thistle Factor

Goldfinches are late nesters compared to many of their avian neighbors. The common wisdom has been that goldfinches wait to nest until thistles go to seed so they can line their nest with the soft seeds. But is this really true? There are other downy seeds available in June, and thistles are a fairly recent arrival to North America. Some theories suggest that July and August nesting may be a response to cowbird nest parasitism or simply that more food is available to feed the nestlings.

Goldfinches line their nest with downy plant fibers and seed fluff.

Nest

Tree- or shrub-nester with nest often placed three to ten feet off the ground. Deep cup that is neatly constructed of plant fibers, bark strips, catkins and lined with downy soft plant seeds. They use spider webs and caterpillar silk to bind the edges. Double brooded (and three broods in a season have been recorded). Usually five eggs but occasionally four or six.

Food

Seeds of weeds, asters and birch; sunflower seeds and niger ("thistle") at feeders.

Nature Notes

State Bird of New Jersey, Washington and Iowa.

Goldfinches are strict vegetarians, surviving and thriving on a diet of seeds. In fact, Brown-headed Cowbird chicks don't survive long in a goldfinch nest as they succumb quickly on a diet lacking insect protein.

4. A summer-plumaged American Goldfinch gets a drink on a hot August day.

Strategies for Winter Warmth

The tiny goldfinch has many winter survival strategies built into its DNA. Like several other species, the goldfinch wings into winter with nearly double the number of feathers it had during the breeding season. And like a good goose-down sleeping bag, more feathers equals more warmth. Fluffing of feathers ("piloerection") while roosting may add 30 to 50 percent insulation value. Roosting in dense cover certainly saves energy, as does dropping its body temperature at night. Shivering, which has a negative connotation in humans, is actually our bodies attempting to generate heat. In goldfinches, it really works and a night of shivering can certainly reduce metabolic costs.

Studies on winter-acclimatized American Goldfinches in the lab found birds subjected to brutal temperatures in the -76° F range were able to maintain normal body temps for six to eight hours by shivering. Heat production was increased by 5.5 times their basal rate just by shivering! The same birds, when captured in summer and subjected to the same extreme cold, would have succumbed in mere minutes. What's the difference? Summer-caught birds run out of carbohydrate reserves quickly and they also don't have enough enzymes to burn fat at a fast enough rate to keep their body temperature up. The winter goldfinches have changed their enzyme balance and can now burn much more fat (which they now, on average, have an extra 2 grams stored for winter) in relation to stored carbs.

Bibliography

Adkisson, Curtis S. 1999. *Pine Grosbeak (Pinicola enucleator), The Birds of North America Online* (A. Poole, Ed.). Ithaca: Cornell Lab of Ornithology; Retrieved from the Birds of North America Online: http://bna.birds.cornell.edu/bna/species/456

Adkisson, Curtis S. 1996. *Red Crossbill (Loxia curvirostra), The Birds of North America Online* (A. Poole, Ed.). Ithaca: Cornell Lab of Ornithology; Retrieved from the Birds of North America Online: http://bna.birds.cornell.edu/bna/species/256

Audubon, Maria R. 1960 (1897). *Audubon and his Journals. Vol. 1. The Labrador Journal.* pp 343-445. Dover Publications, New York, NY.

Badyaev, Alexander V., Virginia Belloni and Geoffrey E. Hill. 2012. *House Finch (Haemorhous mexicanus), The Birds of North America Online* (A. Poole, Ed.). Ithaca: Cornell Lab of Ornithology; Retrieved from the Birds of North America Online: http://bna.birds.cornell.edu/bna/species/046

Baicich, Paul J. and Colin J. O. Harrison. 1997. *A Guide to the Nests, Eggs, and Nestlings of North American Birds* (2nd ed.). Academic Press, San Diego, CA.

Benkman, C. W. 1987. *Food profitability and the foraging ecology of crossbills.* Ecol. Monogr. 57:251-67.

Benkman, C. W. 1988. *Seed handling efficiency, bill structure, and the cost of specialization for crossbills.* Auk 105: 715-19.

Benkman, Craig W. 2012. *White-winged Crossbill (Loxia leucoptera), The Birds of North America Online* (A. Poole, Ed.). Ithaca: Cornell Lab of Ornithology; Retrieved from the Birds of North America Online: http://bna.birds.cornell.edu/bna/species/027

Brinkley, E. S., P. A. Buckley, L. R. Bevier, A. M. Byrne. 2011. *Photo Essay: Redpolls from Nunavut and Greenland visit Ontario.* North American Birds 65, No. 2. 206-215.

Brooks, W. S. 1968. *Comparative adaptations of the Alaskan redpolls to the arctic environment.* Wilson Bull. 80: 253-280.

Brooks, W. S. 1978. *Triphasic feeding behavior and the esophageal diverticulum in redpolls.* Auk 95: 182-183.

Byers, C., J. Curson, & U. Olsson. 1995. *Sparrows and Buntings: A Guide to the Sparrows and Buntings of North America and the World.* Houghton-Mifflin Co. Boston, New York, NY

Cade, T. J. 1953. *Sub-nival feeding of the redpoll in interior Alaska: a possible adaptation to the northern winter.* Condor 55: 43-44.

Carey, C., and R. L. Marsh. 1981. *Shivering finches.* Natural History, 90 (10): 58-63.

Chapman, F. M., and C. A. Reed. 1903. *Color Key to North American Birds.* Doubleday, Page & Co. New York, NY.

Cooper, Jerry. 1995. Birdfinder: *A Birder's Guide to Planning North American Trips.* American Birding Association, Colorado Springs, CO.

Czaplak, D. 1995. *Identifying Common and Hoary Redpolls in winter.* Birding 27: 446-457.

Dawson, W. R., and C. Carey. 1976. *Seasonal acclimatization to temperature in Cardueline finches. I. Insulative and metabolic adjustments.* J. Comparative Physiology, 112: 317-333.

Dawson, William R. 2014. *Pine Siskin (Spinus pinus), The Birds of North America Online* (A. Poole, Ed.). Ithaca: Cornell Lab of Ornithology; Retrieved from the Birds of North America Online: http://bna.birds.cornell.edu/bna/species/280

Dunn, Jon L. and Jonathan Alderfer, eds. 2011. *National Geographic Field Guide to the Birds of North America* (Sixth Edition). National Geographic Society, Washington, DC.

Bibliography

Erskine, A. J. 1977. *Birds in boreal Canada: communities, densities and adaptations*. Can. Wildl. Serv. Rep. Ser. No. 41.

Gillihan, Scott W. and Bruce Byers. 2001. *Evening Grosbeak (Coccothraustes vespertinus), The Birds of North America Online* (A. Poole, Ed.). Ithaca: Cornell Lab of Ornithology; Retrieved from the Birds of North America Online: http://bna.birds.cornell.edu/bna/species/599

Godfrey, W. Earl. 1966. *The Birds of Canada*. The National Museum. Ottawa, ON.

Green, Janet C. 1995. *Birds in Forests: A Management and Conservation Guide*. Minnesota Department of Natural Resources, St. Paul, MN.

Hahn, Thomas Peter. 1996. *Cassin's Finch (Haemorhous cassinii), The Birds of North America Online* (A. Poole, Ed.). Ithaca: Cornell Lab of Ornithology; Retrieved from the Birds of North America Online: http://bna.birds.cornell.edu/bna/species/240

Halfpenny, James C., & Roy Douglas Ozanne. 1989. *Winter: An Ecological Handbook*. Johnson Books, Boulder, CO

Halkin, Sylvia L. and Susan U. Linville. 1999. *Northern Cardinal (Cardinalis cardinalis), The Birds of North America Online* (A. Poole, Ed.). Ithaca: Cornell Lab of Ornithology; Retrieved from the Birds of North America Online: http://bna.birds.cornell.edu/bna/species/440

Harrison, Hal H., 1975. *A Field Guide to the Bird's Nests East of the Mississippi River*. Houghton Mifflin Company, Boston, MA.

Herremans, M. 1990. *Taxonomy and evolution in redpolls Carduelis flammea-hornemanni; a multivariate study of their biometry*. Ardea 78: 441-458.

Johnson, Richard E. 2002. *Black Rosy-Finch (Leucosticte atrata), The Birds of North America Online* (A. Poole, Ed.). Ithaca: Cornell Lab of Ornithology; Retrieved from the Birds of North America Online: http://bna.birds.cornell.edu/bna/species/678

Johnson, Richard E., Paul Hendricks, Donald L. Pattie and Katherine B. Hunter. 2000. *Brown-capped Rosy-Finch (Leucosticte australis), The Birds of North America Online* (A. Poole, Ed.). Ithaca: Cornell Lab of Ornithology; Retrieved from the Birds of North America Online: http://bna.birds.cornell.edu/bna/species/536

Kaufman, Kenn. 2011. *A Field Guide to Advanced Birding*. Houghton Mifflin, Boston, MA.

Knox, Alan G. and Peter E. Lowther. 2000. *Common Redpoll (Acanthis flammea), The Birds of North America Online* (A. Poole, Ed.). Ithaca: Cornell Lab of Ornithology; Retrieved from the Birds of North America Online: http://bna.birds.cornell.edu/bna/species/543

Knox, Alan G. and Peter E. Lowther. 2000. *Hoary Redpoll (Acanthis hornemanni), The Birds of North America Online* (A. Poole, Ed.). Ithaca: Cornell Lab of Ornithology; Retrieved from the Birds of North America Online: http://bna.birds.cornell.edu/bna/species/544

Koenig, W. D., and J. M. H. Knops. *Seed-crop size and eruptions of North American boreal seed-eating birds*. Jrnl. Animal. Ecol. 70: 609-20.

Lyon, B. and R. Montgomerie. 1995. *Snow Bunting (Plectrophenax nivalis)*. In *The Birds of North America*, no. 309 (A. Poole and F. Gill, eds.) Acad. Nat. Sci., Philadelphia, PA and Am. Ornithol. Union, Washington, DC.

Macdougall-Shackleton, Scott A., Richard E. Johnson and Thomas P. Hahn. 2000. *Gray-crowned Rosy-Finch (Leucosticte tephrocotis), The Birds of North America Online* (A. Poole, Ed.). Ithaca: Cornell Lab of Ornithology; Retrieved from the Birds of North America Online: http://bna.birds.cornell.edu/bna/species/559

Marchand, Peter J., 1991. *Life in the Cold: An Introduction to Winter Ecology*. University Press of New England, Hanover, NH.

McEneaney, Terry. 1988. *The Birds of Yellowstone*. Roberts Rinehart, Inc. Publishers, Boulder, CO.

Bibliography

McGraw, Kevin J. and Alex L. Middleton. 2009. *American Goldfinch (Spinus tristis), The Birds of North America Online* (A. Poole, Ed.). Ithaca: Cornell Lab of Ornithology; Retrieved from the Birds of North America Online: http://bna.birds.cornell.edu/bna/species/080

Montgomerie, Robert and Bruce Lyon. 2011. *Snow Bunting (Plectrophenax nivalis), The Birds of North America Online* (A. Poole, Ed.). Ithaca: Cornell Lab of Ornithology; Retrieved from the Birds of North America Online: http://bna.birds.cornell.edu/bna/species/198.

Mullarney, K., L. Svensson, D. Zetterstrom, and P. J. Grant. 1999. *Birds of Europe*. Princeton University Press, Princeton, NJ.

Pielou, E.C. 1988. *The World of Northern Evergreens*. Cornell, Ithaca, NY.

Pulliainen, E. 1978. *The nutritive value of rowan berries, Sorbus aucuparia L., for birds and mammals.* Aquilo Ser. Zool. 18:28-32.

Putnam, L. S. 1949. *The Life history of the Cedar Waxwing.* Wilson Bull. 61: 141-182.

Raynes, Bert, & Darwin Wile. 1994. *Finding the Birds of Jackson Hole*. Darwin Wile. Jackson, WY.

Roberts, Thomas S. 1932. *Birds of Minnesota*. University of Minnesota, Minneapolis, MN.

Roberts, Thomas S., M.D., 1938. *Logbook of Minnesota Bird Life 1917-37*. University of Minnesota Press, Minneapolis, MN.

Sibley, David Allen. 2000. *The Sibley Guide to Birds*. Random House, New York, NY.

Stensaas, Mark. 1993. *Canoe Country Wildlife: A Field Guide to the Boundary Waters and Quetico*. University of Minnesota Press, Minneapolis, MN.

Stephen, L. J. and W. J. Walley. 2000. *Alcohol intoxication contributing to mortality in Bohemian Waxwings and a Pine Grosbeak.* Blue Jay 58: 33-35.

Taylor, P. et. al. 2003. *The Birds of Manitoba*. Manitoba Naturalists Society. Winnipeg, MB.

Tordoff, H. B. 1954. *Social organization and behavior in a flock of captive, nonbreeding Red Crossbills.* Condor 56: 346-358.

Troy, D. M. 1983. *Recaptures of redpolls: movements of an irruptive species.* J. Field Ornithol. 54: 146-151.

Vaughan, Richard. 1992. *In Search of Arctic Birds*. T & AD Poyser Ltd. London.

Witmer, M. C., 1998a. *Ecological and evolutionary implications of energy and protein requirements of avian frugivores eating sugary diets.* Physiol. Zool. 71: 599-610.

Witmer, M. C., D. J. Mountjoy, and L. Elliott. 1997. *Cedar Waxwing (Bombycilla cedrorum).* In *The Birds of North America*, no. 309 (A. Poole and F. Gill, eds.) Acad. Nat. Sci., Philadelphia, PA and Am. Ornithol. Union, Washington, DC.

Witmer, Mark C. 2002. *Bohemian Waxwing (Bombycilla garrulus), The Birds of North America Online* (A. Poole, Ed.). Ithaca: Cornell Lab of Ornithology; Retrieved from the Birds of North America Online: http://bna.birds.cornell.edu/bna/species/714.

Wootton, J. Timothy. 1996. *Purple Finch (Haemorhous purpureus), The Birds of North America Online* (A. Poole, Ed.). Ithaca: Cornell Lab of Ornithology; Retrieved from the Birds of North America Online: http://bna.birds.cornell.edu/bna/species/208.

Websites

www.moumn.org [Minnesota Ornithologist's Union]

www.jeaniron.ca

www.birdingfrontiers.com

www.sibleyguides.com

Photo Credits

The majority of the photos in the book are by the author, Sparky Stensaas. You can see more of Sparky's work at www.ThePhotoNaturalist.com

Paul Bannick (www.paulbannick.com): 23 (sidebar), 35, 36, 37, 40, 41, 46, 52.
Anthony X. Hertzel: 81 (sidebar).
Yann Kolbeinsson (www.pbase.com/birdingiceland/aves): 20 (top), 21 (top left, top right), 70 (sidebar bottom), 75 (sidebar all).
Jason Mandich (www.facebook.com/JasonMandichPhotography): 59, 68 (sidebar left).
Tom Mangelsen (www.mangelsen.com): 47 (bottom), 48.
Andrew Nyhus (www.andrewnyhusphotography.com): 51 (sidebar).
Gerrit Vyn (www.gerritvynphoto.com): 65 (top).
Public Domain/Creative Commons: 25 (sidebar) [courtesy USDA Forest Service via wikimedia.org], 26 (sidebar) [wikimedia.org], 47 (sidebar) [wikipedia.org].

Index

Index

Index

Index